SpringerBriefs in Electrical and Computer Engineering

SpringerBriefs present concise summaries of cutting-edge research and practical applications across a wide spectrum of fields. Featuring compact volumes of 50 to 125 pages, the series covers a range of content from professional to academic. Typical topics might include: timely report of state-of-the art analytical techniques, a bridge between new research results, as published in journal articles, and a contextual literature review, a snapshot of a hot or emerging topic, an in-depth case study or clinical example and a presentation of core concepts that students must understand in order to make independent contributions.

More information about this series at http://www.springer.com/series/10059

Saleh Faruque

Radio Frequency
Propagation Made Easy

 Springer

Saleh Faruque
Department of Electrical Engineering
University of North Dakota
Grand Forks
North Dakota
USA

ISSN 2191-8112   ISSN 2191-8120 (electronic)
SpringerBriefs in Electrical and Computer Engineering
ISBN 978-3-319-11393-7   ISBN 978-3-319-11394-4 (eBook)
DOI 10.1007/978-3-319-11394-4

Library of Congress Control Number: 2014954243

Springer Cham Heidelberg New York Dordrecht London
© Springer International Publishing Switzerland 2015

Printed on acid-free paper

Springer is part of Springer Science+Business Media (www.springer.com)

- It is shown that free space pathloss exhibits an equation of a straight line, having a pathloss slope of 2.
- We have defined the Effective Radiated Power (ERP) and the Received Signal Level (RSL) and have shown that RSL exhibits an equation of a straight line having a slope of -2.

In Chap. 3:

- We have examined the Fresnel Zone Effects and various anomalies of RF propagation and have shown that there exists a free space propagation medium in multipath environments.
- We have presented a two ray model for outdoor deployment and have shown that these propagation models also exhibit equation of straight line within the Fresnel zone break point.
- We have also presented a two ray model for indoor deployment and a multrti ray propagation model
- For tunnels and subways and have shown that these propagation models also exhibit equation of straight line within the Fresnel zone break point.
- These findings indicate that cellular network based on Fresnel zone break point as the cell radii is an effective solution to reduce power and save energy.

In Chap. 4:

- We have presented a general overview of various empirical prediction models and have shown that these propagation models also exhibit equation of straight line within the Fresnel zone break point.
- Although these predictions and measurement techniques are the foundation of today's cellular services, they suffer from inaccuracies due to user defined clutter factors. These clutter factors arise due to numerous RF barriers which vary from place to place. It is practically impossible to accommodate all these factors accurately. Cell site location is also a challenging engineering task because of regulations and restrictions imposed on some locations. Therefore cell sites have to be relocated from the predicted location, requiring best judgment of RF engineers. Thus we came to the conclusion that propagation prediction is a combination of science, engineering and art. An experienced RF engineer, willing to compromise between theory and practice, is expected to accomplish the most.

In Chap. 5:

- We have reviewed Statistical Analysis and showed that it is an important exercise to design and implement cellular base stations with reliability.
- Presented regression analysis and showed that random data such as Received Signal Level (RSL) can be predicted with confidence.
- Drive Test, Live air data collection & Data analysis techniques were presented.
- A computer aided prediction technique was presented as a student project.

Preface

Hello! Do you recognize me? I am the cell phone, sometimes also called the palm phone or mobile phone. Many of you have grown familiar with using me; but have you ever given a thought to what I am, and how I became so intimately involved in your world of communications? The story of my invention sounds almost like a fairy tale to today's new generation. I was invented in the USA, in the eighties of the twentieth century. The history of my advent is crowded with momentous events (see Chap. 1). The scientific community believes that I came into being due to light experiments, which in turn led to the development of electromagnetic theory, followed by radio wave propagation and then to my being developed as a major communication device. I communicate through Radio frequency (RF), which is the rate of oscillation in the range of around 800 MHz–5.6 GHz. This brings us to the salient concept of the this book entitled "Radio Frequency Propagation Made Easy".

Radio frequency propagation made easy is a booklet that brings you up-to-date in key concepts, underlying principles and practical applications of wireless communications. This book has seven chapters comprising various aspects of radio wave propagation and its attributes. The contents of these chapters are briefly presented below:

In Chap. 1:

- We have traced the historical background and have shown that the modern wireless communication system is due to a series of light experiments.
- The mechanism and the underlying principle of electromagnetic radiation were presented with illustrations.
- The key concepts, underlying principles and construction of array antennas were provided.
- This material was very lucidly and simply presented, so as to be easy for readers to grasp.

In Chap. 2:

- We have derived the free-space path loss formula and have shown that it is *proportional* to the square of the distance.
- Free space pathloss is also proportional to the square of the *frequency*.

In Chap. 6:

- We have discussed radio Frequency coverage and provided the concept of cell
- Rationalized the use of hexagonal cell geometry and calculated cell radius
- Provided the concept of OMNI and Sectorized cells
- Provided the concept of Cell cluster
- Presented $N = 7$ frequency reuse plan and carrier to interference ratio (C/I)
- Presented $N = 3$ frequency reuse plan and carrier to interference ratio (C/I)
- Discussed the benefit of antenna down tilt and calculated the down tilt angle

In Chap. 7:

- We have discussed Global RF & CO_2 Pollution connected to wireless communications.
- The classical Electron Spin Resonance (ESR) is presented to show that there is a possible
- public health issues due to RF absorption.
- It has also been argued that cell phone technology may contribute to global CO_2 pollution, expected to rise due to high speed data communication.
- With this in mind, we have presented a technique to design energy efficient green cellular technology, comprising Micro, Pico and Femto cells.

My readers, in this book I've tried to present radio frequency propagation facts for you in easy language. If this book pleases you while it teaches, I shall be amply rewarded.

Contents

Chapter 1
Introduction to Radio Frequency

Objectives:

- Trace the history
- Show the relationship of light, electricity and magnetism
- Present the key concepts and underlying principles of electromagnetic radiation.
- Provide the basic understanding of array antenna.

1.1 Tracing the History

The radio wave, also known as Electro-Magnetic (EM) wave, is invisible to the human eye and has a speed of 186,000 miles per second. Many of us have grown familiar with using EM waves; but have we ever given a thought to what the EM wave is, and how it became so intimately involved in our world of communications? The story of its presence will sound almost like a fairy tale to today's new generation. The history of its advent is crowded with momentous events. The scientific community believes that light and EM wave are closely related. Here is a brief history:

Perhaps it all began with a single command "Let there be light" and there was light (Genesis 1:3), It took 13.5 billion years, homo-erectus stood up, saw the light and began wondering, what's all this? This insatiable desire to know the unknown and see the unseen has propelled mankind since time immemorial to ever newer inventions.

As civilization progressed, man attained the power to make various sounds. These sounds first combined to form coherent words; these words were then used to make simple sentences, which gave rise to the spoken language. To disseminate the language among the masses came the written word, and a new way of communication began.

It took another millennium to determine that White light is a mixture of lights of seven colors: red, orange, yellow, green, blue, indigo and violet (Newton-1680) [1, 2]. See Fig. 1.1. When white light consisting of seven colors falls on a transparent medium (glass prism), each color in it is refracted (or deviated) by a different angle,

© Springer International Publishing Switzerland 2015
S. Faruque, *Radio Frequency Propagation Made Easy*, SpringerBriefs in Electrical and Computer Engineering, DOI 10.1007/978-3-319-11394-4_1

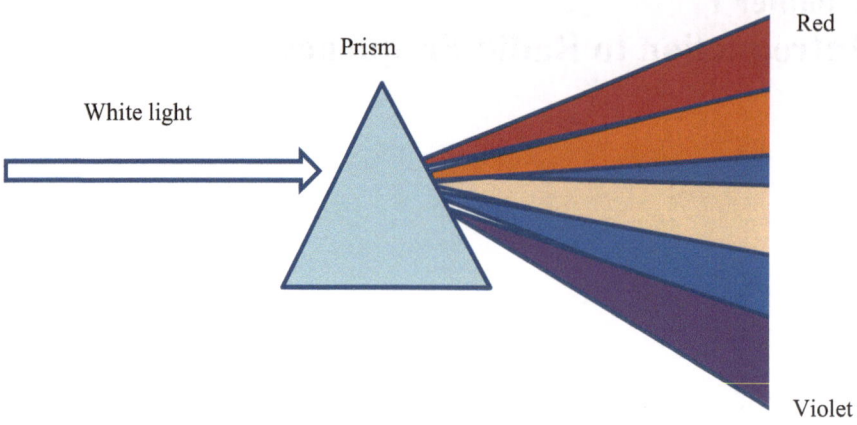

Fig. 1.1 Light has seven colors. (Newton 1680)

with the result that seven colors are spread out to form a spectrum, The red color is deviated least, so it forms the upper part of the spectrum. On the other hand, Violet color deviates the most.

   This discovery led to a series of light experiments, in turn leading to several significant developments, notably that light, electricity and magnetism are related (Michael Faraday-1831) [3]. In 1864 Maxwell proved that Faraday had been right and that light is indeed electromagnetic and its velocity is 186,000 miles per second [4]. These parameters are related by the following formula:

$$c = f \lambda \tag{1.1}$$

Where $c = 3 \times 10^8$ m/s is the velocity of light, f = frequency and λ = wavelength.

Problem 1.1 Given: f = 1 MHz, $c = 3 \times 10^8$ m/s. Find λ.

Solution 1.1

$$\lambda = c/f$$

$$= (3 \times 10^8 \text{ m/s})/(1 \times 10^6 \text{ Hz})$$

$$= 300 \text{ m}$$

Problem 1.2 Given: f = 1 GHz, $c = 3 \times 10^8$ m/s. Find λ.

Solution 1.2

$$\lambda = c/f$$

$$= (3 \times 10^8 \text{ m/s})/(1 \times 109 \text{ Hz})$$

$$= 0.3 \text{ m}$$

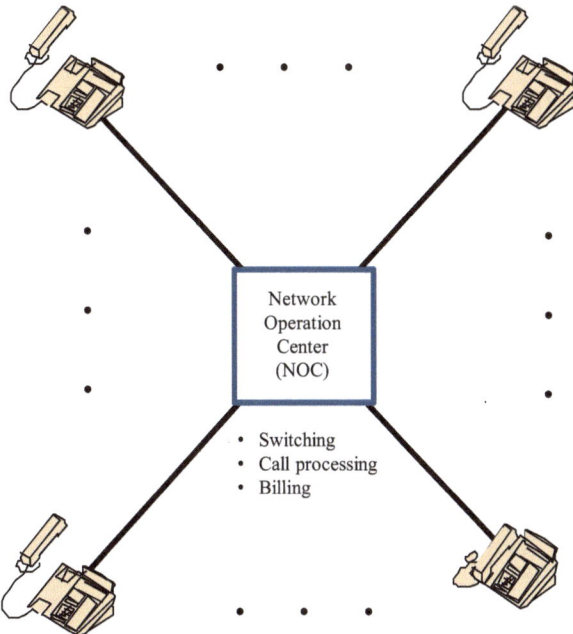

Fig. 1.2 The telephone system. Tens of thousands of telephone lines are connected to a switching center. The switching center also known as the Network Operation Center (NOC), which provides the bill

From the two problems above, we find that the wavelength is inversely proportional to the frequency, where the velocity of light is the proportionality constant. This formula is a very significant in antenna design, as we shall see later in this chapter.

Now let's go back to the history. Since the urge to see the unseen, know the unknown and explore the unexplored has always propelled man, he did not stop here. The intense desire to send the sound of the human voice far away through wires prompted Scottish physicist Alexander Graham Bell to invent the telephone [5]. Today, the telephone system can support tens of thousands of telephone customers through the telephone switching center shown in Fig. 1.2.

However, would the ever-curious human mind be satisfied with just this? Of course not. So now the world scientific community began racking its brains to come up with a way to transmit sound wirelessly. After years of careful planning and much intense research activity, the Italian physicist Tomas Guglielmo Marconi invented the radio in 1897 (Fig. 1.3) [6, 7]. For this invention, he was awarded the Nobel Prize for Physics in 1929.

A short while after Marconi invented the radio, man found out that the radio transmits sound only. Why, he could not see the picture of the speaker on it! As a result, now the scientific community busied itself in trying to come up with a device, which would broadcast sounds and pictures together. As they thought, so they did. American scientists invented the television (Fig. 1.4) [8, 9]. Today, the television is in millions of homes across the world.

Radio Transmitter

Fig. 1.3 The radio, 1897

TV Antenna

Fig. 1.4 The television

But man is never content just to sit. Yet it would also be impossible to take the telephone or the television along when one went out. Thus now the scientific community directed its efforts towards a device which would enable the user to talk as well as watch television while on the move; thus the mobile phone, also known as cell phone, was born [10–14]. Today, cell-phone technology is spreading like wildfire across the world. Figure 1.5 shows the modern cellular communication network, providing wireline and wireless links to users. Typically, cell phone towers are installed in different propagation environments as follows:

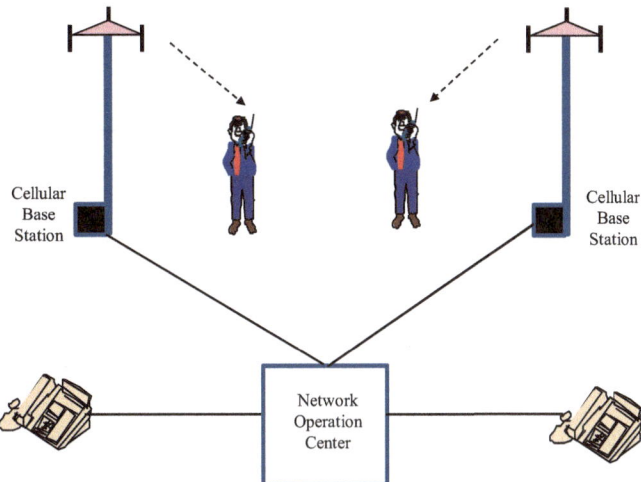

Fig. 1.5 Cellular communication network

- Every 0.5–1 mile in dense urban environments
- Every 1–3 miles in urban environments
- Every 3–10 miles in suburban environments and
- Every 10–30 miles in rural environments.

The density of these cell phone towers depend on the population density. Tens of thousands of users are supported by these cellular networks in a typical metropolitan area.

Notice that, the traditional land telephone network is also connected to the cellular network via the NOC (Network Operation Center). The NOC provides connectivity, call processing, billing etc.

The table below summarizes all the significant developments due to light experiments including the invention of the radio (Table 1.1).

Table 1.1 Significant developments due to light experiment

What	Who	When
Seven colors of light	Newton	1680
Infrared ray	Herschel	1800
Ultraviolet ray	Ritter	1801
Wave theory of light	Thomas Young	1805
Proof of wave theory	Fresnel	1805
Relationship between light, electricity and magnetism	Faraday	1831
Electromagnetic theory	Maxwell	1864
Proved Maxwell's theory by Leyden Jar experiment	Hertz	1866
Radio	Marconi	1897
Television	RCA, USA	1955
Cellular telephone	Bell Labs, USA	Late 1980s

Table 1.2 The electromagnetic spectrum

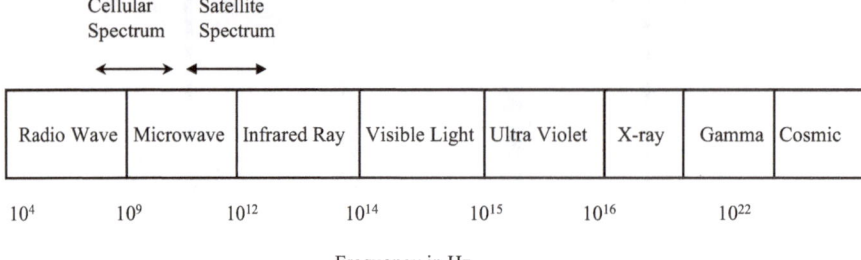

Frequency in Hz

Today, we have AM radio, FM radio, Television, Satellite, Cordless phones, Cell phones, blue Tooth, Wi-Fi, WiMAX, etc. These consumer products are available all over the world and they use different band of frequencies. Table 1.2 shows the entire electromagnetic spectrum. These frequency components differ in energy depending on the wavelength. Their ability of propagation is different in different medium. All electromagnetic waves are Transverse waves. Notice that only a fraction of this spectrum is used for wireless communications.

Table 1.3 shows the range of frequencies used by wireless Systems.

Table 1.3 Frequency bands for wireless communications

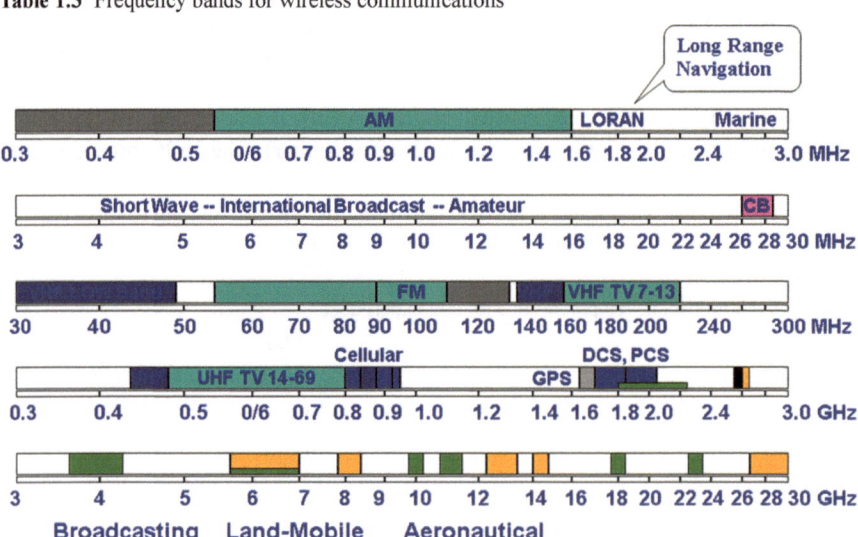

1.2 Electromagnetic Wave

To conclude this introductory chapter, a brief overview of the mechanism of electromagnetic radiation is presented, followed by antenna technology [15–18].

1.2.1 The Electric Field

Electromagnetic wave consists of both electric and magnetic fields. The electric field E is due to voltage and the magnetic field is due to current. Let's consider the electric field first as shown in Fig. 1.6. In Fig. 1.6a, we have an open circuit, excited by a d.c source voltage. Because the circuit is open, there is no current flow through the circuit, i.e., the current I = 0. However, there is an electric field E due to the voltage. The intensity of the electric field E decays as a function of distance. Notice that the direction of electric field is from the positive terminal to the negative terminal of the battery as shown in Fig. 1.6a. Also note that there is no magnetic field, since the current is zero (I = 0).

In Fig. 1.6b, we have another open circuit, with the polarity of the voltage reversed. In this circuit, we see that the direction of E field also changes in accordance with the polarity of the voltage, while the current I = 0 as well. From the above, we see that the direction of E field changes in accordance with the polarity of the d.c. voltage.

1.2.2 Electric and Magnetic Fields

The electric field E is due to voltage and the magnetic field H is due to current. We examine this by means of an open circuit, excited by an a.c. source, as shown in Fig. 1.7. The operation of the circuit is as follows:

- During the positive half cycle of the input signal, there exists an electric field E due to voltage and a magnetic field H due to current. The electric field E is as-

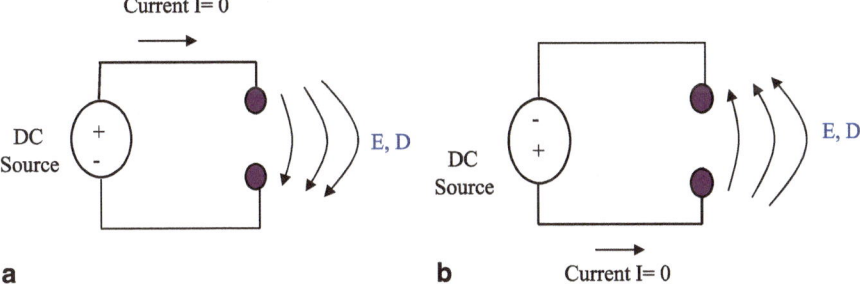

Fig. 1.6 Electric field intensity E due to D.C. source. The electric field E is associated with an electric flux density D. Direction of E field depends on the polarity of the voltage

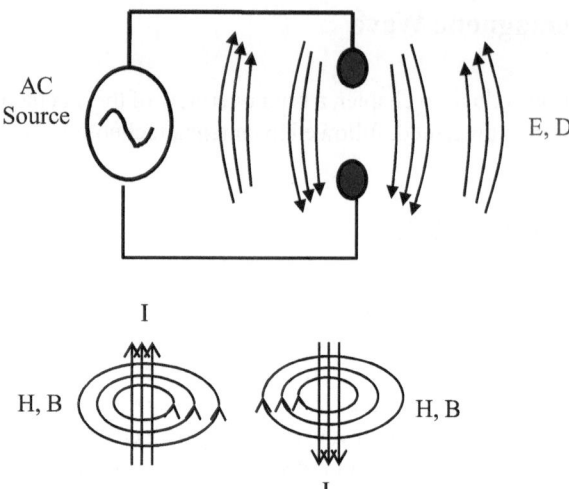

Fig. 1.7 Electric and magnetic field intensity at a distance from an a.c. source. The direction of E and H field changes in accordance with the polarity of the source voltage. The electric field E is associated with an electric flux density D and the magnetic field H is associated with a magnetic flux density B. The E field and the H field are perpendicular to each other

sociated with an electric flux density D and the magnetic field H is associated with a magnetic flux density B. See Fig. 1.7 for illustrations. The intensity and direction of the E and H fields follow in accordance with the intensity and polarity of the a.c. source. The E field and the H field are perpendicular to each other.

- Similarly, during the negative half cycle of the input signal, there exists an electric field E due to voltage and a magnetic field H due to current. The electric field E is associated with an electric flux density D and the magnetic field H is associated with a magnetic flux density B. See Fig. 1.7 for illustrations. The intensity and direction of the E and H fields follow in accordance with the intensity and polarity of the a.c. source. The E field and the H field are perpendicular to each other.

According to Maxwell [4]:

- The electric flux density D is proportional to the electric field intensity E.
  i.e., $D = \varepsilon E$, where ε is the proportionality constant.
  ($\varepsilon = 8.854 \times 10^{-12}$ C^2/Nm^2)
- The magnetic flux density B is proportional to magnetic field intensity H.
  i.e., $B = \mu H$, where μ is the proportionality constant.
  ($\mu = 4\ \mathrm{pi} \times 10^{-7}$ Wb/Am)
- The velocity of light c in free space is:
  $c = (\varepsilon\mu)^{-1/2} = 3 \times 10^8$ m/s (186,000 miles per second)

1.2.3 Mechanism of Electro Magnetic Radiation

Electromagnetic radiation has two components: (a) An electric field E and (b) a magnetic field H, where the E field is due to voltage and the H field is due to current. These E and H fields are orthogonal (perpendicular) to each other. According to Maxwell's electromagnetic theory [4], the velocity of electromagnetic wave is 186,000 miles per second (3×10^8 m/s). Figure 1.8 illustrates the mechanism of electromagnetic radiation. Let's take a closer look:

- During the positive half cycle of the input voltage, there is a buildup of E field and H field. The intensity and the direction of E and H fields are in accordance with the polarity of the voltage as shown in Fig. 1.8.
- When the polarity of the voltage changes (negative half cycle), the E and H fields already developed earlier forms a loop and are released from the dipole, allowing the negative field to build up (see Fig. 1.8). The condition for this release is $\lambda/2$, where λ is the wavelength.
- This process continues for subsequent cycles. This is the well-known electromagnetic wave, which can be viewed as a concentric sphere expanding at the speed of light, as shown in Fig. 1.8.

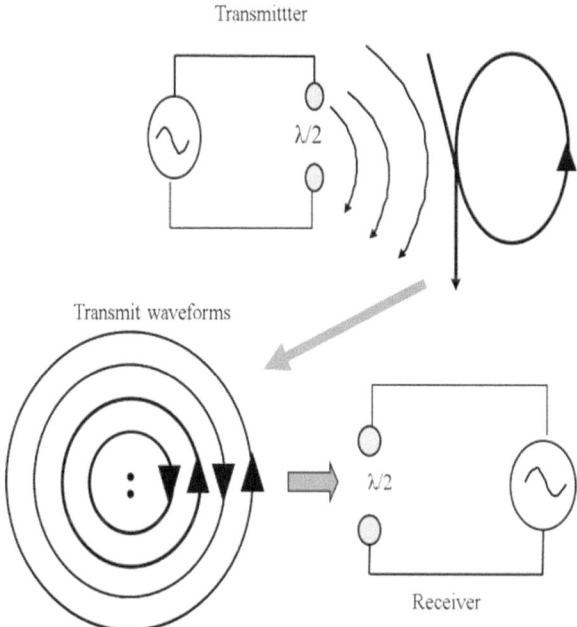

Fig. 1.8 Illustration of Electromagnetic radiation

1.2.4   How to Intercept Electro Magnetic Radiation

Figure 1.8 also shows how to intercept the radiated electromagnetic signal by means of an identical dipole. Because of the alternating nature of the electromagnetic waves, a dipole having a gap λ/2 will induce the signal which has been transmitted by a transmit dipole. This instrument is called a dipole antenna. Therefore an antenna is also a reciprocal device.

1.3   Antenna Basics

1.3.1   Dipole Antenna

An antenna is a reciprocal device which means that it transmits and receives electromagnetic waves. We examine this by means of a single dipole, as shown in Fig. 1.9a. The outcome is a radiation pattern which radiates in all directions. The radiation pattern is known as OMNI (all) directions.

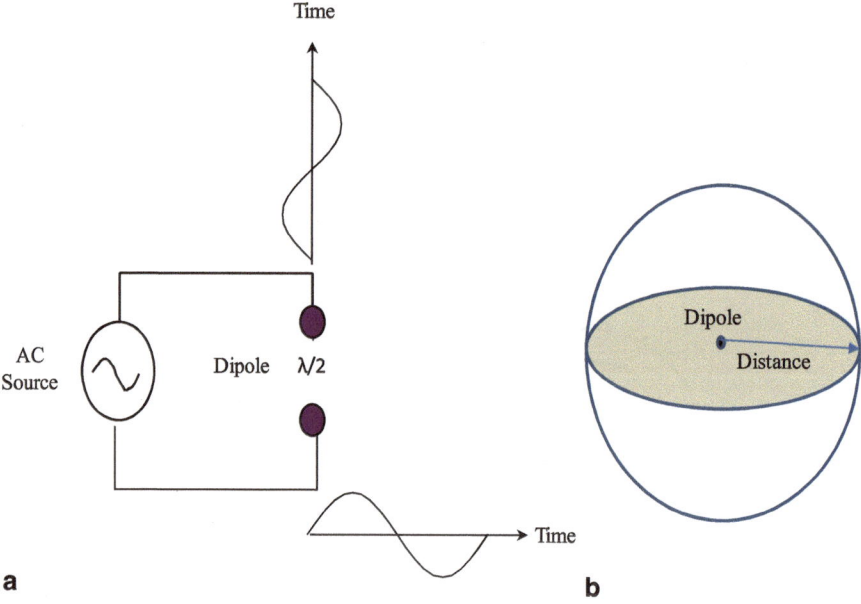

Fig. 1.9 a Single dipole. **b** Radiation in all directions

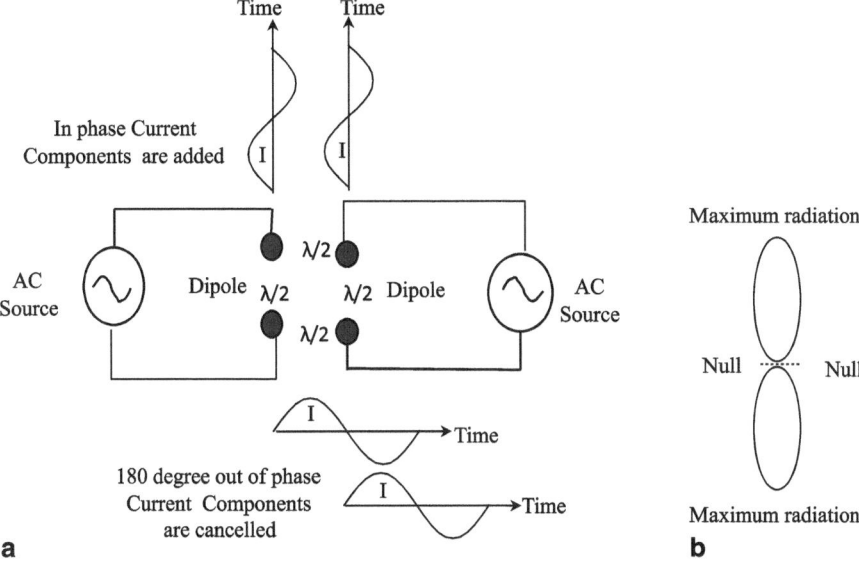

Fig. 1.10 Array antenna. **a** Two linearly excited dipole antenna. **b** Radiation pattern

1.3.2 Uniformly Excited Linear Array Antenna

An antenna using multiple radiating dipoles is known as array antenna. Figure 1.10a shows two linearly excited arrays, where:

- In-phase components in the vertical plane are added to provide a gain of two (3dB)
- The 180°. out of phase components in the horizontal plane are cancelled to provide a null.
- The outcome is a radiation pattern which has a gain of 2 (3 dB) in the vertical plane. See Fig. 1.10b.

We can create different radiation pattern by changing the phase. Figure 1.11 shows an example of two linearly excited arrays, having a phase difference of 180°. Here we see that,

- In-phase components in the horizontal plane are added to provide a gain of two (3 dB)
- The 180° out of phase components in the vertical plane are cancelled to provide a null.
- The outcome is a radiation pattern which has a gain of 2 (3 dB) in the horizontal plane. See Fig. 1.11b

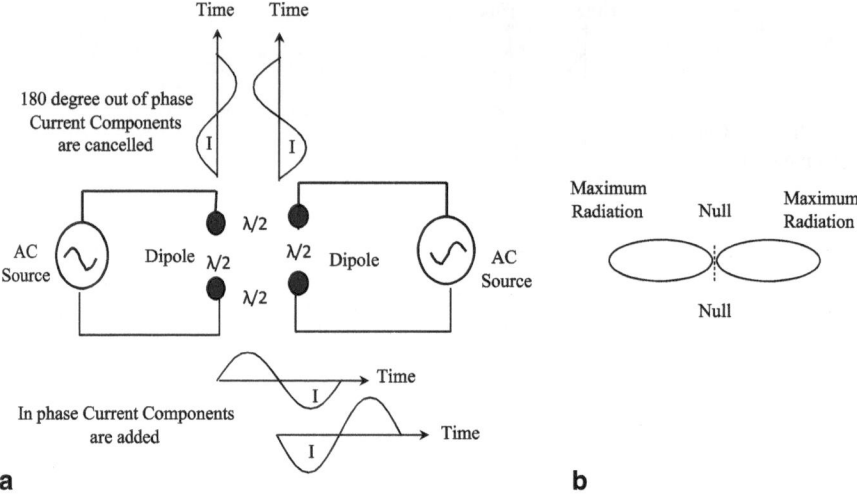

Fig. 1.11 Array antenna. **a** Two linearly excited dipole antenna. **b** Radiation pattern

1.3.3   Non-uniformly Excited Linear Array Antenna

An antenna using multiple radiating dipole elements, excited by different voltages or currents, is known as non-uniformly excited array antenna. Figure 1.12 shows an example of two non-linearly excited arrays, where:

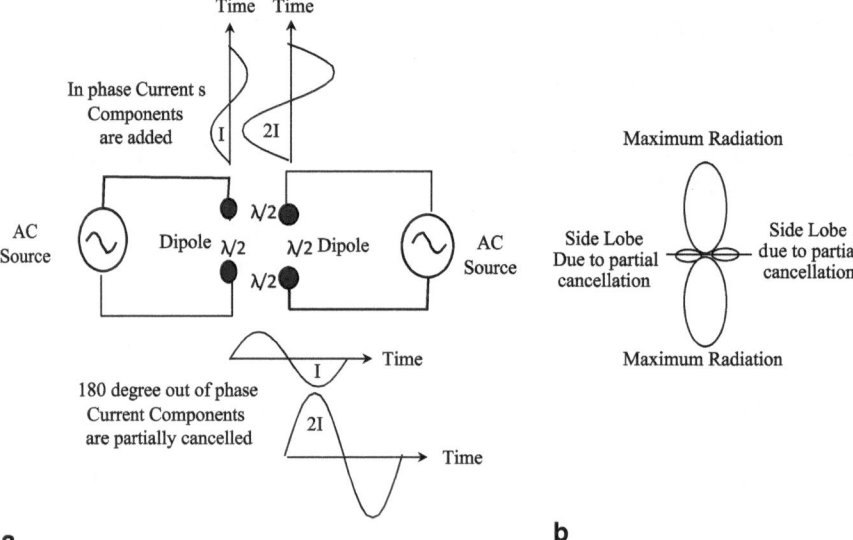

Fig. 1.12 **a** Non-Uniformly excited array antenna having two radiating elements and its **b** Radiation pattern

- In-phase components in the vertical plane are added to provide a gain in the main lobe.
- The 180° out of phase components in the horizontal plane are partially cancelled to provide a side lobe.
- The outcome is a radiation pattern which has main lobes in the vertical plane and side lobes in the horizontal plane as shown in Fig. 1.12.

1.3.4 Side Lobe Cancellation in Non-uniformly Excited Linear Array Antennas

We can also cancel side lobes by creating different currents and phases in array antennas. Figure 1.13 shows an example of three non-linearly excited arrays, having different currents and phases, while cancelling the side lobes in the horizontal plane. Thus referring to Fig. 1.13, we see that,

- In the vertical plane, in-phase current components are added to provide a composite gain of $1+2+1=4$, where
  $I_1 = 1$, having no delays in the vertical plane
  $I_2 = 2$, having no delays in the vertical plane
  $I_3 = 1$, having no delays in the vertical plane

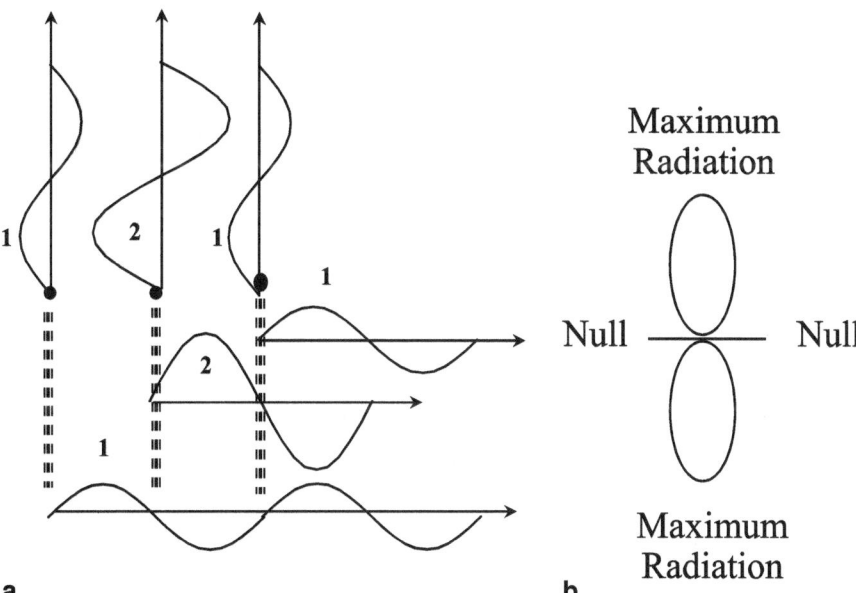

Fig. 1.13 a Non-Uniformly excited array antenna having three radiating elements and its **b** Radiation pattern

- In the horizontal plane, the current components are cancelled as $1-2+1=0$, where
  $I_1=1$, having a delay of 360° in the horizontal plane
  $I_2=2$, having a delay of 180° in the horizontal plane
  $I_3=1$, having no delays in the horizontal plane
- The outcome is a radiation pattern which has main lobes in the vertical plane and nulls in the horizontal plane.as shown in Fig. 1.13b.

1.3.5 Radiation Patterns of Commercial Antennas

Typical commercial cellular antennas are vertical combinations of dipoles. The horizontal plane pattern is determined by the number of horizontally-spaced elements. The vertical plane pattern is determined by number of vertically-separated elements. The radiation patterns of these antennas are usually plotted in polar form as shown in Fig. 1.14. The horizontal plane radiation pattern is a function of azimuth. The vertical plane radiation pattern is a function of elevation.

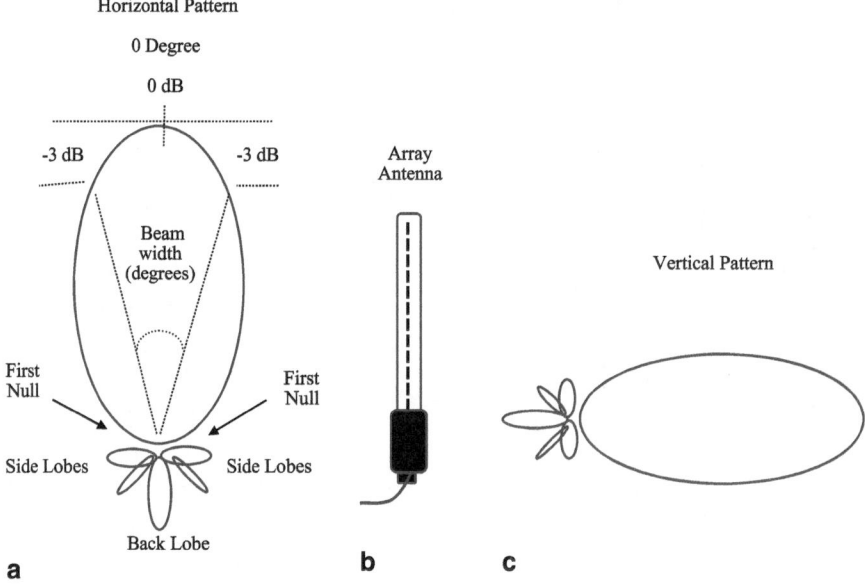

Fig. 1.14 a Horizontal pattern. **b** Array antenna and **c** Vertical pattern

Antennas are often compared by noting specific features on their patterns, such as,

- − 3 db. − 6 dB, − 10 dB points
- Antenna gain
- Antenna beam width
- Angles of null
- Front-to-back ratio
- Frequency response and bandwidth
- Main lobe, side lobes, back lobe, etc.

Some of these points are briefly presented below [19]:

1.3.6 Antenna Gain

Antenna gain is defined with respect to isotropic gain. It determines the degree of energy concentration in one direction with respect to other directions as shown in Fig. 1.15. Note that the total energy is constant, i.e., the total energy within the main lobe plus the side lobes is the same as in the isotropic region.

Fig. 1.15 Antenna gain

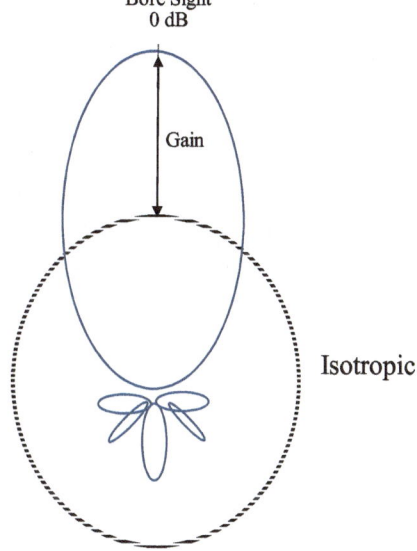

Fig. 1.16 Antenna beam
width

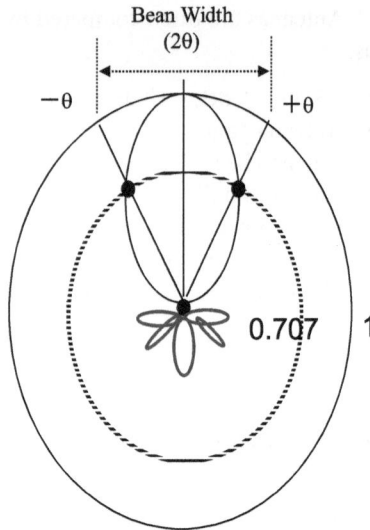

1.3.7 Antenna Beam Width

Antenna beam width is defined as 2θ, where θ is the angle with respect to the bore sight where the voltage is 0.707 of its maximum value. This is shown in Fig. 1.16.

1.3.8 Front To Back Ratio

Antenna front-to-back ratio is defined as the ratio of the power radiated from the main lobe to that of the back lobe. See Fig. 1.17 for illustration. It is an important design parameter in cellular communications, because of frequency reuse and C/I (Carrier to Interference) constraints. We shall address this topic again in the future in this book.

Antenna front to back ratio is defined as:

$$\text{Front - to - Back Ratio} = 10\text{Log}_{10}\left[\frac{P_{\text{main lobe}}}{P_{\text{back lobe}}}\right] \qquad (1.2)$$

1.3.9 Frequency Response and Bandwidth

Every antenna has a frequency response. It passes certain frequencies and blocks other frequencies. See Fig. 1.18. The bandwidth is given by:

$$\text{BW} = (f_{\text{H}} - f_{\text{L}}) \qquad (1.3)$$

Fig. 1.17 Front to back ratio

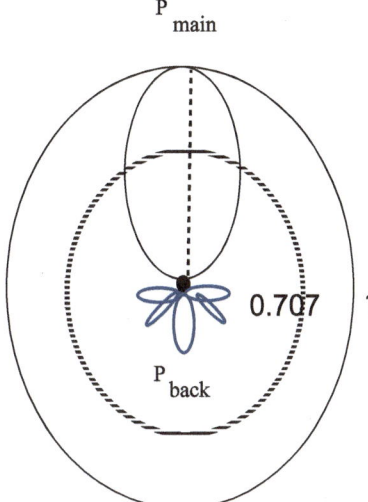

Fig. 1.18 Frequency response of antenna

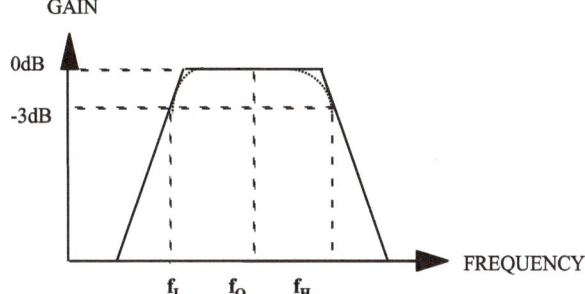

Where,

f_H = Upper 3-dB frequency
f_L = Lower 3-dB frequency
f_0 = Center frequency

1.4   Conclusions

- We have traced the historical background and have shown that the modern wire-less communication system is due to a series of light experiments.
- The mechanism and the underlying principle of electromagnetic radiation were presented with illustrations.
- The key concepts, underlying principles and construction of array antennas were provided.
- This material was very lucidly and simply presented, so as to be easy for readers to grasp.

References

1. John C. D. Brand (1995). Lines of light: the sources of dispersive spectroscopy, 1800–193019301800–1930. CRC Press. pp. 30–323230–32. ISBN 978-2-88449-163-1.
2. Mary Jo Nye (editor) (2003). The Cambridge History of Science: The Modern Physical and Mathematical Sciences 5. Cambridge University Press. p. 278. ISBN 978-0-521-57199-9.
3. Michael Faraday entry at the 1911 Encyclopedia Britannica hosted by LovetoKnow January 2007.
4. The Scientific Papers of James Clerk Maxwell Volume 1 page 360; Courier Dover 2003, ISBN 0-486-49560-4
5. Alexander Graham Bell, ":Encyclopedia Britannica", 2009.
6. Guglielmo Marconi, British patent No. 12,039 (1897) "Improvements in Transmitting Electrical impulses and Signals, and in Apparatus therefor". Date of Application 2 June 1896; Complete Specification Left, 2 March 1897; Accepted, 2 July 1897 (later claimed by Oliver Lodge to contain his own ideas which he failed to patent).
7. Guglielmo Marconi, British patent No. 7,777 (1900) "Improvements in Apparatus for Wireless Telegraphy". Date of Application 26 April 1900; Complete Specification Left, 25 February 1901; Accepted, 13 April 1901.
8. United States Patent Office, Patent No. 2,133,123, 11 Oct 1938.
9. United States Patent Office, Patent No. 2,158,259, 16 May 1939
10. "1946: First Mobile Telephone Call". corp.att.com. AT&T Intellectual
11. 1947 memo by Douglas H. Ring proposing hexagonal cells. (PDF). Retrieved on 2012-12-30.
12. "Switching Plan for a Cellular Mobile Telephone System":, Z. Fluhr and E. Nussbaum, IEEE Transactions on Communications volume 21, #11 p. 1281 (1973)
13. Hachenburg, V.; Holm, B.D.; Smith, J.I. (1977). "Data signaling functions for a cellular mobile telephone system". IEEE Transactions on Vehicular Technology 26: 82. doi:10.1109/T-VT.1977.23660.
14. Martin Cooper, et al., "Radio Telephone System", US Patent number 3,906,166; Filing date: 17 October 1973; Issue date: September 1975; Assignee Motorola.
15. "Applications of electromagnetic induction". Boston University. 1999-07-22.
16. Sadiku, M. N. O. (2007). Elements of Electromagnetics (fourth ed.). New York (USA)/Oxford (UK): Oxford University Press. p. 386. ISBN 0-19-530048-3.
17. Ulaby, Fawwaz (2007). Fundamentals of applied electromagnetics (5th ed.). Pearson: Prentice Hall. p. 255. ISBN 0-13-241326-4.
18. J.D. Kraus, Antenna, McGraw-Hill, New York, 1960.
19. Saleh Faruque, "Cellular Mobile Systems Engineering", Artech House, Norwood, MA, ISBN: 0-89006-518-7, 1996.

Chapter 2
Free Space Propagation

Objectives

- Define free space propagation in time and space
- Derive free space pathloss formula
- Show that free space pathloss formula exhibits an equation of straight line
- Define ERP and RSL
- Problems
- Group exercise

2.1 Free-Space Propagation in Time and Space

Electromagnetic waves differ in energy according to their wavelength. Their ability to propagate is also different in different propagation environments. In free space (vacuum) they are characterized by their ability to propagate without obstruction and without atmospheric effects. The path loss under these conditions is said to be free space path loss.

For example, we consider an isotropic RF (Radio Frequency) source, which radiates electromagnetic energy uniformly in all directions, as shown in Fig. 2.1a in three-dimensional space. The radiating source is located at the center, which begins its emission at a given time. Maxwell's theory of Electromagnetic Radiation implies that the energy radiates uniformly in all directions, at the speed of light (3×10^8 m/s or 3.3 µs/km) [1]. This may be viewed as a sphere, expanding in time and space.

Since it is difficult to represent time and space in four dimensions, we can represent this time-space relationship by means of a cross-sectional view of the energy sphere in two dimensions in space and one dimension in time [2]. This is shown in Fig. 2.1b, where time is represented in the vertical axis.

There is no signal outside the cone, since the velocity of electromagnetic wave is constant. For example, we assume that $d_1 = 1$ km, $d_2 = 2$ km and $d_3 = 3$ km, the RF signal that originates at time $t = 0$, will arrive in those locations exactly after 3.3, 6.6 and 9.9 µs respectively. This implies that the propagated signal exists within

© Springer International Publishing Switzerland 2015
S. Faruque, *Radio Frequency Propagation Made Easy,* SpringerBriefs in Electrical and Computer Engineering, DOI 10.1007/978-3-319-11394-4_2

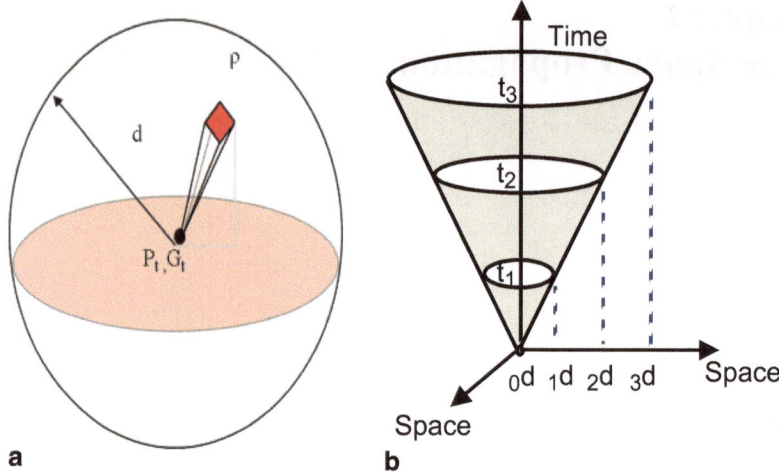

Fig. 2.1 Illustration of free Space propagation in time and space. (**a**) An isotropic RF (Radio Frequency) source, which radiates electromagnetic energy uniformly in all directions. (**b**) Representation of time-space relationship by means of a cross-sectional view of the energy sphere in two dimensions in space and one dimension in time

a space-time coordinate (d_i, t_i) where d_i is the location of the signal and ti is the corresponding instant of time. The propagation delay is given by

$$t_p \approx 3.3 \ \mu \text{s/km} \tag{2.1}$$

This propagation delay is an important parameter in cellular communication systems. It determines the maximum cell size and inter symbol interference in digital cellular radios.

For example the signal that arrives at d_3, has a propagation delay of 9.9 μs. This delay determines the radius of the sphere, which has uniform signal strength throughout the surface of the sphere. In cellular communication this sphere is known as an ideal cell, as shown in Fig. 2.1a. The total energy within the cell is constant irrespective of time and space.

2.2 Derivation of Free Space Pathloss Formula

Consider the free space propagation model as shown in Fig. 2.1a to derive the well-known free space pathloss formula [3]. Assuming that the total transmit power at the source as P_t, whose gain in a particular direction is G_t, and then the radiated power density at a given distance d will be given by

$$\rho = \frac{P_t G_t}{4\pi d^2} \ watts \ / \ m^2 \tag{2.2}$$

If a receive antenna is located at a distance d, whose gain is G_r and the effective area is A [2], where

$$A = G_r \frac{\lambda^2}{4\pi} \tag{2.3}$$

The received power P_r at the terminal of the receive antenna will be given by

$$P_r = \rho A = P_t G_t G_r \left(\frac{\lambda}{4\pi d}\right)^2 \tag{2.4}$$

In the above analysis, it was assumed that the transmission began at t_0 and that it was received at a distance d at t. The time difference $t - t_0$ is the propagation delay (t_p), which is given by,

$$t_p = t - t_0 \tag{2.5}$$

Thus, by knowing the start time, the propagation delay and hence the distance can be determined.

Now, referring to Eq. 2.4 we find that the received signal attenuates as square of the distance. The pathloss formula is given by the ratio of the received power to the transmit power, i.e.,

$$L_p = \left(\frac{P_r}{P_t G_t G_r}\right) \tag{2.6}$$

Combining Eqs. 2.4 and 2.6, we get,

$$L_p = \left(\frac{\lambda}{4\pi d}\right)^2 \tag{2.7}$$

In decibel, the free space path loss formula (Lp), can be obtained as

$$L_P(dB) = 10 \log \left[\left(\frac{4\pi d}{\lambda}\right)^2\right] \tag{2.8}$$

or

$$L_p(dB) = 32.5 + 20 \log(f) + 20 \log(d) \tag{2.9}$$

Where $\lambda = c/f$, $c = 3 \times 10^8$ m/s, the frequency (f) is measured in MHz and the distance (d) is measured in km. Equation (2.9) is the familiar free space path loss formula.

2.3   Free Space Path Loss Formula Exhibits Equation of Straight Line [4]

Consider the free space path loss formula again as given below:

$$L_p(dB) = 32.5 + 20\log(f) + 20\log(d)$$

For a given frequency, 20 log(f) is constant. Therefore, we can express the above equation as

$$L_p(dB) = L_o(dB) + \gamma 10\log(d) \qquad (2.10)$$

Notice that, Eq. (2.10) is similar to the equation of straight line of the form:

$$y = c + mx \qquad (2.11)$$

Where,

$y = L_p$ in dB
$c = L_o(dB) = 32.5 + 20\log(f)$ is the intercept in dB
$m = \gamma = 2$ is the slope
$x = 10\log(d)$ is the distance in logarithmic scale

Figure 2.2 Shows the path loss characteristics for a given frequency.

   From the above analysis, we see that free space propagation exhibits an equation of a straight line having a pathloss slope of 2 ($\gamma = 2$). The intercept L_o depends on the frequency. Later in this book we shall see that all propagation models can be approximated as an equation of a straight line having different pathloss slopes, depending on the propagation environment.

Problem 2.1

Given:
- Frequency $f = 1\,GHz\,(10^9\,Hz)$
- Distance $d = 10\,km$
- Free space Path loss slope $\gamma = 2$

Find:

a. The intercept L_o in dB
b. The free space pathloss L_p in dB

Fig. 2.2 Free space path loss characteristics in a log-log scale. The path loss slope $\gamma = 2$

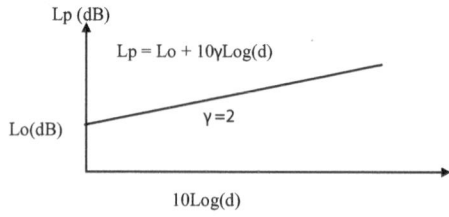

Lp (dB)

Lp = Lo + 10γLog(d)

Lo(dB)

γ = 2

10Log(d)

Solution:

a. Intercept: $L_o\,(\text{dB}) = 32.5 + 20\text{Log}\,(1000\text{ MHz})$

$$= 32.5 + 60$$
$$= 92.5\text{ dB}$$

b. Path loss $L_p\,(\text{dB}) = L_o\,(\text{dB}) + \gamma 10\text{Log}\,(\text{d})$

$$= 98.5 + 10 \times 2 \times 10\text{Log}\,(10\text{ km})$$
$$= 98.5 + 20 \times 1$$
$$= 118.5\text{ dB}$$

Problem 2.2

Given:

*   Frequency f $= 2$ GHz
*   Distance d $= 10$ km
*   Free space Path loss slope $\gamma = 2$

Find:

a. The intercept L_o in dB
b. The free space path loss L_p in dB

Solution:

a. Intercept: $L_o\,(\text{dB}) = 32.5 + 20\text{Log}\,(2000\text{ MHz})$

$$= 32.5 + 66$$
$$= 98.5\,\text{dB}$$

c. Path loss $L_p\,(\text{dB}) = L_o\,(\text{dB}) + \gamma\,10\text{Log}\,(\text{d})$

$$= 98.5 + 10 \times 2 \times 10\text{Log}\,(10\text{ km})$$
$$= 98.5 + 20 \times 1$$
$$= 118.5\text{ dB}$$

From the above two problems, we see that, for a given distance, the intercept and pathloss increase by 6 dB when the frequency doubles.

2.4   ERP and RSL

The Effective Radiated Power (ERP) and the Received Signal Level (RSL) are two design parameters used in cellular communications. ERP is the power radiated from the tip of the antenna and RSL is the power received at the receiver. The receiver is located at a distance d from the transmitter. We examine this by means of Fig. 2.3:

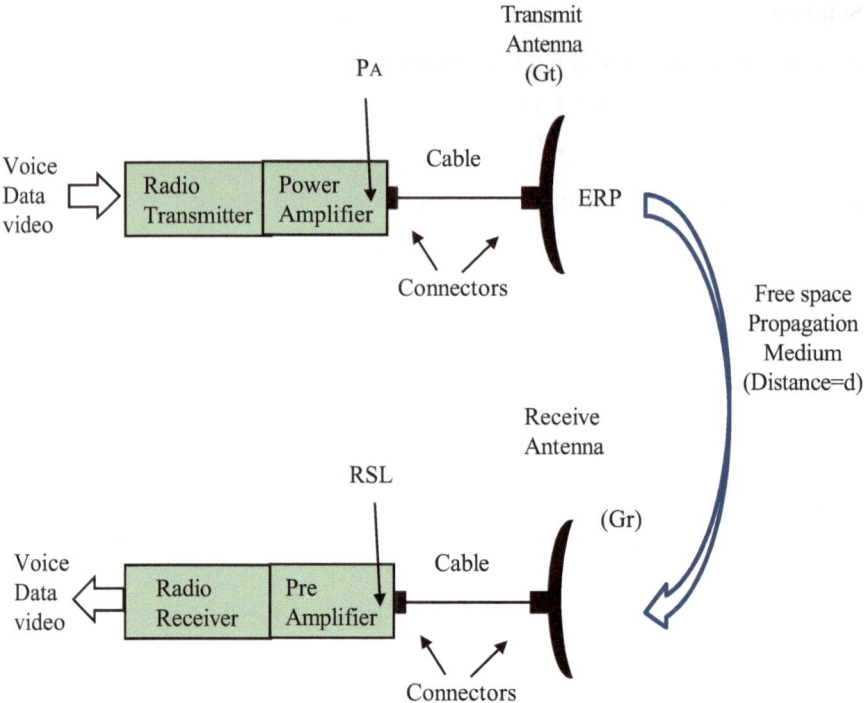

Fig. 2.3 A free space radio link. ERP is the power radiated from the tip of the antenna and RSL is the power received at the receiver. The transmitter and the receiver are separated by a distance d

In Fig. 2.3, let the effective radiated power be defined as ERP and the received signal level be defined as RSL. Then, we can write:

$$ERP = P_A - 2L_{connector} - L_{cable} + G_t \qquad (2.12)$$

$$RSL = ERP - L_p + G_r - L_{cable-2Lconnector} \qquad (2.13)$$

Where,

- P_A is the output power from the power amplifier
- $L_{connector}$ is the connector loss. The factor 2 is due to two connectors
- L_{cable} is the cable loss
- G_t is the transmit antenna gain
- G_r is the receive antenna gain
- L_p is the pathloss

The effective radiated power ERP is constant since all the parameters in Eq. 2.12 are constants. Now, Substituting for L_p, in Eq. 2.13 we get,

$$RSL = ERP - 32.5 - 20Log(f) - 20Log(d) + G_r - L_{cable} - 2L_{connector}$$

Fig. 2.4 Illustration of path-loss slope and RSL

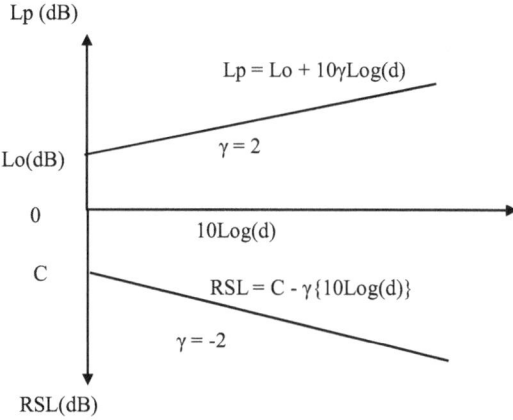

Since ERP, frequency, G_r, Cable loss and Connector losses are constant parameters, the above equation reduces to,

$$RSL = C - 20Log(d)$$
$$= C - \gamma 10Log(d) \qquad (2.14)$$

Where,

$$C = ERP - 32.5 - 20Log(f) + G_r - L_{cable} - 2L_{connector} \qquad (2.15)$$

Notice that the received signal level RSL also exhibits an equation of a straight line having an intercept C and a slope $\gamma = -2$. This is plotted in Fig. 2.4 along with the path loss characteristic.

Problem 2.3

Consider a radio link as shown in Fig. 2.3 with the following design parameters:

Frequency	$f = 1\,GHz$
Propagation medium	Free space
Distance	$d = 10\,km$
Power from the amplifier	$P_A = 10$ watts
Transmit cable and connector losses	3 dB
Transmit antenna gain	$G_t = 10\,dB$
Receiver antenna gain	$G_r = 10\,dB$
Receive cable and connector losses	3 dB

Find:

a. ERP in dB
b. Path loss in dB
c. Received Signal Level RSL in dBm

Solution:

a. $\text{ERP(dB)} = 10\text{Log}\left(P_A\right) - \text{Cable \& Connector losses} + G_t$

$\qquad = 10\text{ dB} - 3\text{ dB} + 10\text{ dB}$

$\qquad = 17\text{ dBW}$

b. $L_p = 32.5 + 20\text{Log (1000 MHz)} + 20\text{Log (10 km)}$

$\qquad = 32.5 + 60 + 20$

$\qquad = 112.5\text{ dB}$

c. $\text{RSL} = \text{ERP} - L_p + G_r - \text{Cable \& Connector Losses}$

$\qquad = 17\text{ dB} - 112.5\text{ dB} + 10\text{ dB} - 3\text{ dB}$

$\qquad = 27\text{ dB} - 115.5\text{ dB}$

$\qquad = -88.5\text{ dB}$

$\qquad = -88.5 + 30$

$\qquad = -58.5\text{ dBm}$

[Note: 1 W=0 dBW. Also, 1 W=1000 mW=30 dBm. Therefore, 0 dBW=30 dBm]

2.5   Conclusions

• We have derived the free-space path loss formula and have shown that it is proportional to the square of the distance.
• Free space pathloss is also proportional to the square of the frequency.
• It is shown that free space pathloss exhibits an equation of a straight line, having a pathloss slope of 2.

References

1. The Scientific Papers of James Clerk Maxwell Volume 1 page 360; Courier Dover 2003, ISBN 0-486-49560-4
2. Stephen Hawking, "A Brief History of Time", Bantam Books, New York, 1988.
3. J.D. Kraus, Antenna, McGraw-Hill, New York, 1960
4. Saleh Faruque, "Cellular Mobile Systems Engineering", Artech House Inc. ISBN: 0-89006-518-7, 1996.

Chapter 3
Multipath Propagation

Objectives

- Define multipath propagation
- Review Fresnel Zones and show that there exists a free space propagation medium in multipath environments.
- Examine a two ray propagation model and show that there exists a free space propagation medium in the terrestrial environment.
- Examine a three ray propagation model for indoor applications and show that there exists a free space propagation medium. In indoor environment.
- Examine a multi ray propagation model for tunnels and subways and show that there exists a free space propagation medium.

3.1 Introduction to Multipath

Multipath propagation is due to reflection, diffraction and scattering of radio waves caused by obstructions along the path of transmission as shown in Fig. 3.1.

The magnitude of these effects depends on the type and total area of obstruction. For example, a plane surface of vast area will produce maximum reflection while a sharp object such as a mountain peak or an edge of a building will produce scattering components with minimum effects known as knife-edge effect. These spurious signals have longer path lengths than the direct signal. The associated magnitude and phase differences also vary according to the path length. We examined these by means of Fresnel Zone effects.

© Springer International Publishing Switzerland 2015   27
S. Faruque, *Radio Frequency Propagation Made Easy,* SpringerBriefs in Electrical
and Computer Engineering, DOI 10.1007/978-3-319-11394-4_3

Fig. 3.1 Illustration of
a multipath propagation
environment

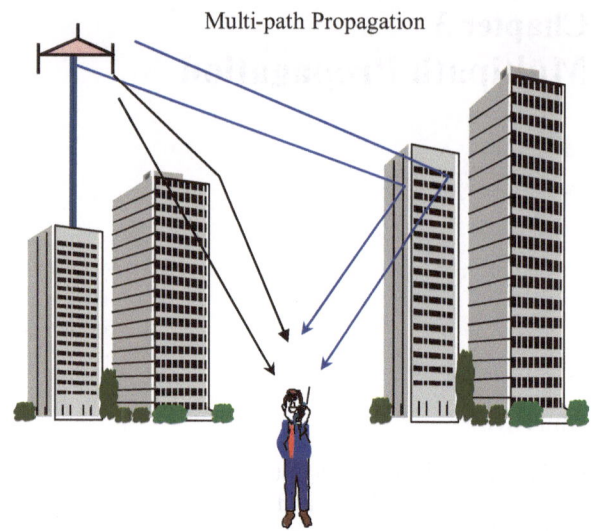

Multi-path Propagation

3.2   Effect of Fresnel Zone

In multipath environments, diffraction of radio waves occurs when the wave front encounters an obstacle. A model originally developed by A. Fresnel for optics can examine this [1]. Fresnel postulated that the cross section of an optical wave front (electromagnetic wave front) is divided into zones of concentric circles, separated by $\lambda/2$ (Fig. 3.2) where λ is the wavelength.

The radius of the nth Fresnel zone is given by

$$R_n = \left[n\, \lambda\, (d_1\, d_2\, /\, (d_1 + d_2)) \right]^{1/2} \tag{3.1}$$

Where

d_1 distance between the transmitter and the obstruction
d_2 distance between the receiver and the obstruction
λ c/f
n   1 for the first Fresnel zone
n   2 for the second Fresnel zone

Fig. 3.2 Illustration of
Fresnel zone

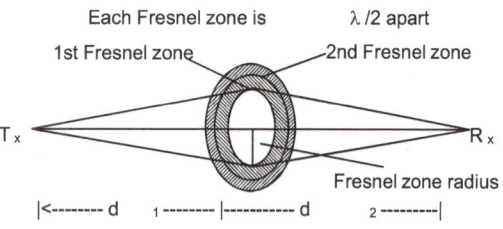

Each Fresnel zone is λ /2 apart

1st Fresnel zone   2nd Fresnel zone

T_x R_x

Fresnel zone radius

|<-------- d $_1$ -------- |----------- d $_2$ ---------|

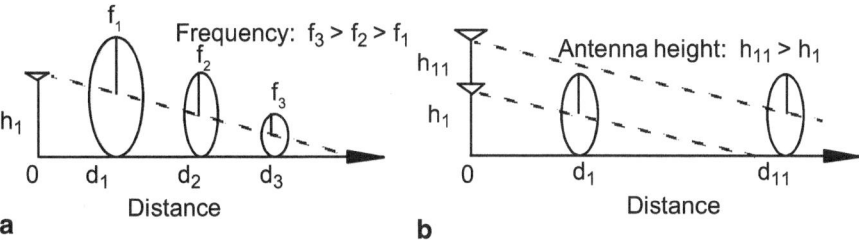

Fig. 3.3   a A high frequency signal propagates further before the first Fresnel zone touches the ground. **b** A signal radiating from a tall antenna propagates further before the first Fresnel zone touches the ground

From Eq. 3.1, we find that the Fresnel zone radius is inversely proportional to the square root of frequency. This implies that for a given antenna height, a high frequency signal will propagate further before the first Fresnel zone touches the ground (Fig. 3.2a). Likewise, for a given frequency, a signal that radiates from a tall antenna will propagate further before the first zone touches the ground (Fig. 3.2b). In other words, diffraction of radio waves depends on frequency as well as on antenna height. We examine this by means of a two ray model as follows:

3.3   The Existence of a Free Space Propagation Medium in Outdoor Propagation Environment

We begin our investigation by considering an outdoor propagation medium having a flat terrain, as shown in Fig. 3.3 [2, 3]. Here,

h_1 Transmit (Tx) antenna height
h_2 Receive (Rx) antenna height
d   Antenna separation
d1   Length of the reflected path
d2   Length of the direct path.

From plane geometry, the path differences between the direct and the reflected path can be estimated as,

$$\Delta d = \left[(h_1 + h_2)^2 + d^2 \right]^{1/2} - \left[(h_1 - h_2)^2 + d^2 \right]^{1/2} \tag{3.2}$$

Where $\Delta d = d2 - d1$. After some algebraic manipulation, Eq. 6.35 may be expressed as,

$$\Delta d = \frac{4h_1 h_2}{d \left[\left\{ (h_1 + h_2) / d \right\}^2 + 1 \right]^{1/2} + d \left[\left\{ (h_1 - h_2) / d \right\}^2 + 1 \right]^{1/2}} \tag{3.3}$$

Fig. 3.4 Two-ray model

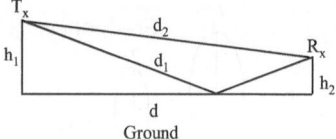

Where $(h1 \pm h2)/d \ll 1$ and the path difference reduces to

$$\Delta d \approx 2h_1h_2/d \qquad (3.4)$$

When the path difference between the direct ray and the diffracted ray is $\lambda/2$, diffraction will be maximum. Thus from Eq. 3.4 we write,

$$\Delta d \approx 2h_1h_2/d = \lambda/2 \qquad (3.5)$$

Resulting in

$$d_o = d = 4h_1h_2/\lambda = 4h_1h_2f/c \qquad (3.6)$$

Where

f   frequency
c   velocity of light

This distance (d_o) is known as the Fresnel zone break point, which is proportional to frequency and antenna height as shown in Fig. 3.4. The line of sight pathloss slope within d_o is similar to free space pathloss since diffraction and multipath phenomenon generally occurs beyond this region (Fig. 3.5).

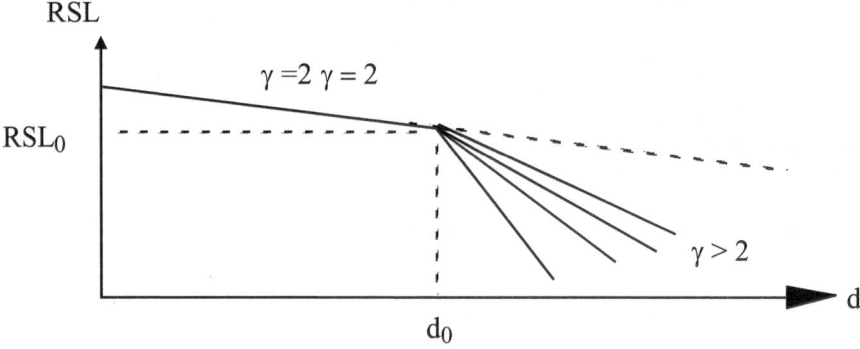

Fig. 3.5 Illustration of Fresnel zone break point

3.4 An Alternate Proof

The analysis presented in the preceding section can be verified by combining the powers received from the direct path and the received path. Thus, referring to the two-ray model, the composite received power can be expressed as [4]

$$P_r = P_t \left(\lambda/4 \pi d \right)^2 \cdot \left[1 + e^{-j\Delta\phi} \right]^2 = \left[P_t \left(\lambda/4 \pi d \right)^2 \right] \cdot \left[4\mathrm{Sin}^2 \left(\Delta\phi/2 \right) \right] \qquad (3.7)$$

Where $\Delta\phi$ is the phase difference between the direct and the reflected path. In terms of path difference, it is given by

$$\Delta\phi = (2\pi/\lambda)\Delta d \qquad (3.8)$$

Combining (3.7) and (3.8) we obtain

$$P_r = \left[P_t (\lambda/4\pi d)^2 \right] \cdot \left[4\sin^2 (\pi/\lambda)\Delta d \right] \qquad (3.9)$$

And with $\Delta d \approx 2h1h2/d$, we get

$$P_r = \left[P_t (\lambda/4\pi d)^2 \right] \cdot \left[4\sin^2 \left\{ (\pi/\lambda).2h_1 h_2/d \right\} \right] \qquad (3.10)$$

Which is maximum for

$$(\pi/\lambda) \cdot 2h_1 h_2/d\} = (\pi/2) \qquad (3.11)$$

or

$$d_o = d = 4h_1 h_2/\lambda \qquad (3.12)$$

Under this condition, the received power can be obtained with the following constraints:

1. Antenna separation: $d >> h_1$ and h_2
2. Incident angle is negligible
3. Phase difference ($\Delta\phi$) is negligibly small.

Then Eq. (3.10) reduces to

$$P_r \approx \left[P_t (\lambda/4\pi d_o)^2 \right] \qquad (3.13)$$

Which is the free space loss? It may be noted that d is now replaced by d_o, where d_o is the Fresnel zone break point. Therefore, path loss characteristics within d_o will be similar to free space path loss, i.e. sq. law attenuation. The signal attenuates faster beyond d_o due to destructive multipath components, represented by Eq. 3.10 and plotted in Fig. 3.6.

From the preceding analysis we conclude that there is a free space path loss region before the Fresnel zone break point. After the break point, the signal attenuates faster depending on the propagation medium. This break point is a function of Frequency and transmit/receive antenna heights.

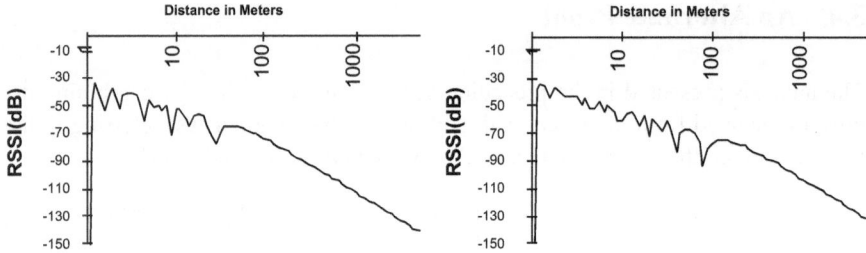

Fig. 3.6 Received signal level as a function of distance showing the break point. There exists a free space propagation environment before the break point

Problem 3.1 Given:

- Frequency = 900 and 1900 MHz
- Tx antenna height = 30 m
- Rx antenna height = 1.5 m

Compute the break point (d_o).

Answer

$do = 4 \times 30 \times 1.5 \times 900 \times 10^6 / 3 \times 10^8$

$\quad = 640$ m @ 900 MHz

$do = 4 \times 30 \times 1.5 \times 1900 \times 10^6 / 3 \times 10^8$

$\quad = 1140$ m @ 1900 MHz

3.5   The Existence of a Free Space Propagation Medium in Buildings and Shopping Malls

In-building coverage is a major concern among service providers mainly because of high attenuation within the building. We examine this by means of a three-ray model, where the antenna is located within the building having ground reflections as well as reflections from the ceiling [3]. This is shown in Fig. 3.7 where

H	Ceiling height
h_1	Transmit antenna height
h_2	Receive antenna height
d	Antenna separation
D	Direct path
d_1	Ground reflected path
d_2	Ceiling reflected path

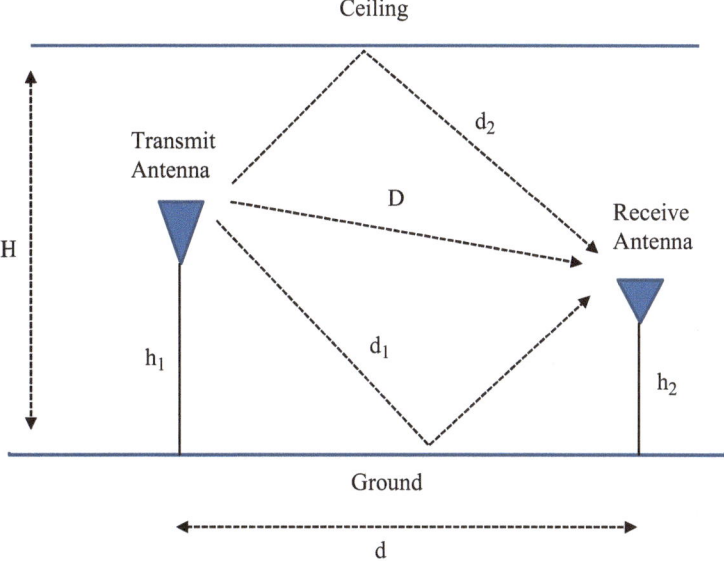

Fig. 3.7 Three-ray indoor model

Because there are two dominant reflections, one from the floor and the other from the ceiling, our objective is to make these path differences identical so that the first Fresnel zone break point occurs at the same point. It should be noted that reflections from sidewalls and their path differences are unpredictable since the portable cellular phone moves horizontally. Moreover, these horizontal components suffer from numerous losses due to penetration through glass, porous materials etc. before reflection. For these reasons, we ignore the horizontal components from the following analysis.

From plane geometry, the path differences between the direct and the vertical reflected paths (Fig. 6.19) can be estimated as,

$$\Delta d_1 = \left[(h_1 + h_2)^2 + d^2 \right]^{1/2} - \left[(h_1 - h_2)^2 + d^2 \right]^{1/2} \tag{3.14}$$

$$\Delta d_2 = \left[\{(H-h_1)+(H-h_2)\}^2 + d^2 \right]^{1/2} - \left[\{(H-h_1)-(H-h_2)\}^2 + d^2 \right]^{1/2} \tag{3.15}$$

Where $\Delta d1 = d1 - D$ and $\Delta d2 = d2 - D$

After some algebraic manipulation, Eq. 6.28 and Eq. 6.29 may be expressed as,

$$\Delta d_1 = \frac{4h_1 h_2}{d\left[\{(h_1 + h_2)/d\}^2 + 1 \right]^{1/2} + d\left[\{(h_1 - h_2)/d\}^2 + 1 \right]^{1/2}} \tag{3.16}$$

$$\Delta d_2 = \frac{4(H-h_1)(H-h_2)}{d\left[\{(2H-h_1 - h_2)/d\}^2 + 1 \right]^{1/2} + d\left[\{(h_2 - h_1)/d\}^2 + 1 \right]^{1/2}} \tag{3.17}$$

Where $(h_1 \pm h_2)/d \ll 1$ and $2H - h_1 - h_2 \ll 1$. Thus we approximate,

$$\Delta d_1 \approx 2 h_1 h_2 / d \qquad (3.18)$$

$$\Delta d_2 \approx 2(H - h_1)(H - h_2)/d \qquad (3.19)$$

Since it is desirable to have identical path differences, we write

$$\Delta d_1 = \Delta d_2 = \Delta d \qquad (3.20)$$

For which we obtain the following identity:

$$H = h_1 + h_2 \qquad (3.21)$$

The composite path difference then appear as

$$\Delta d_2 = 2(H - h_2) h_2 / d \qquad (3.22)$$

which is a function of the ceiling height and the receive antenna height. Within the first Fresnel zone, this path difference is exactly $\lambda/2$, for which there is maximum diffraction. Thus from Eq. 6.36 we write,

$$2(H - h_2) h_2 / d = \lambda / 2 \qquad (3.23)$$

Resulting in [3.10]

$$d_o = d = 4(H - h_2) h_2 / \lambda \quad \text{(Indoor)} \qquad (3.24)$$

Therefore, for optimum performance, the base station antenna should be located below the ceiling by h2 where h2 is the portable antenna height with respect to the floor. In other words, base station antenna height with respect to the ceiling = portable antenna height with respect to the floor.

Problem 3.2 Given:

- Frequency = 900 MHz and 1900 MHz
- Ceiling height = 4.5 m
- Rx antenna height = 1.5 m

Compute the break point (do)

Answer:

Tx antenna height = $4.5 - 1.5 = 3$m
$d_o = 4(H - h_2) h_2 / \lambda$
$d_o = 4 \times 3 \times 1.5 \times 900 \times 10^6 / 3 \times 10^8$
$\quad = 64$ m @ 900 MHz
$d_o = 4 \times 3 \times 1.5 \times 1900 \times 10^6 / 3 \times 10^8$
$\quad = 114$ m @ 1900 MHz

3.6 The Existence of a Free Space Propagation Medium in Tunnels and Subways

In this section we examine multi-ray propagation in tunnels and subways. It is shown that for a given frequency and device geometry, there exists a Fresnel Zone break point, where the direct path and the reflected path are exactly 180° out of phase. RF signals within this distance are in-phase and do not cancel each other out. On the other hand, RF signals beyond this point are out of phase and suffer from multipath cancellations. Therefore, by knowing the frequency and the geometry of the device, the RF transmit and receive antennas can be properly positioned to create a free-space propagation region between the antennas [3].

To illustrate the concept, let's consider RF propagation inside an open ended circular metallic tube as shown in Fig. 3.8, where the antenna separation is less than the Fresnel Zone break point. As a result, multipath components will be cancelled and the propagation will be similar to free space propagation within d_0.

The tube has the following design parameters:

D   Diameter of the tube
h_1 Transmit antenna height
h_2 Receive antenna height
d_0 Antenna separation
d_3 Direct path
d_1 Reflected path-1
d_2 Reflected path-2

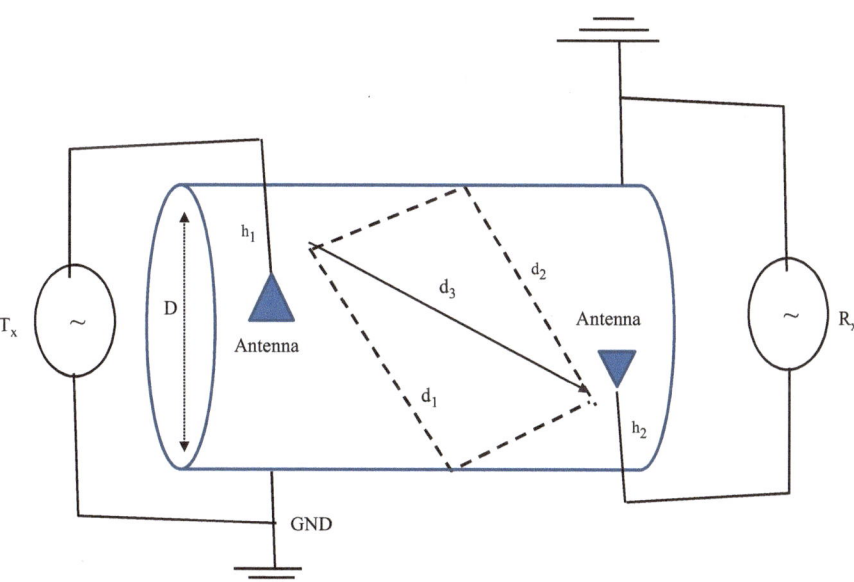

Fig. 3.8 RF propagation in tunnels and subways

It is assumed that for every reflected path there is an identical path from the opposite wall inside the cylindrical tube. So that the path differences are identical, i.e.,

$$\Delta d = d_1 - d = d_2 - d \tag{3.25}$$

Using plane geometry, it can be shown that,

$$\Delta d = 2(D - h_1)(D - h_2)/d \tag{3.26}$$

The maximum diffraction occurs when the path difference between the direct ray and the diffracted ray is $\lambda/2$. Therefore eq. 3.2 can be written as,

$$\lambda/2 = 2(D - h_1)(D - h_2)/d \tag{3.27}$$

or

$$d = d_o = 4(D - h_1)(D - h_2)/\lambda \tag{3.28}$$

Where d_o is the antenna separation for which the diffraction is maximum. It follows that for a distance $d < d_o$, the phase difference between the direct and the reflected path will be less than 180° and they will not cancel each other out. Consequently, the pathloss slope within d_o (Fresnel Zone Break Point) will be similar to free space path loss, i.e. sq. law attenuation. The signal will attenuate faster beyond do due to diffractions and multipath cancellations.

The total received power can be estimated by combining the powers received from the direct path and the reflected path [4]. Thus, referring to the three-ray model, the composite received power can be expressed as:

$$P_r = P_T \left(\lambda/4\pi d_o \right)^2 \left[1 + e^{-j\Delta\phi} \right]^2 = \left[P_T (\lambda/4\pi d_o)^2 \right] \cdot \left[4\sin^2(\Delta\phi/2) \right] \tag{3.29}$$

Where P_T is the transmit power, d_o is the Fresnel zone break point and $\Delta\phi$ is the phase difference between the direct and the reflected path. In terms of path difference, it is given by

$$\Delta\phi = (2\pi/\lambda)\,\Delta d_o \tag{3.30}$$

Combining (3.29) and (3.30) we obtain

$$P_r = \left[P_T (\lambda/4\pi d_o)^2 \right] \cdot \left[4\sin^2(\pi/\lambda)\Delta d_o \right] \tag{3.31}$$

The path difference Δd_o is given by Eq. 3.2. Substituting Eq. 3.2 into Eq. 3.7, we obtain,

$$P_r = \left[P_T + (\lambda/4\pi d_o)^2 \right] \left[4\sin^2(\pi/\lambda)\{2(D - h_1)(D - h_2/d_o)\} \right] \tag{3.32}$$

We also note that in Eq. 3.8, the maximum received power is achieved when,

$$(\pi/\lambda)\, 2\, (D1 - h_1)(D2 - h_2)/d_o\} = \pi/2 \tag{3.33}$$

Combining Eq. 3.8 and Eq. 3.9, we get,

$$P_r \approx \left[P_T(\lambda/4\pi d_o)^2\right] \tag{3.34}$$

Where P_T is the transmit power, P_r is the receive power, λ is the wavelength, and d_o is the distance, representing the Fresnel zone break point. The pathloss formula within the tube is then,

$$L_p = P_r/P_T = (\lambda/4\pi d_o)^2 \tag{3.35}$$

Which is the familiar free space pathloss formula? Therefore we can conclude that there exists a free-space pathloss within a circular tube, provided the antenna separation is less than the Fresnel zone break point d_o. The equation for d_o is presented again or convenience;

$$d = d_o = 4(D - h_1)(D - h_2)/\lambda \tag{3.36}$$

Problem 3.3 This problem relates to cellular antenna deployment in a subway tunnel. The tunnel has the following design parameters:

- D = 10 m (Diameter of the tunnel)
- h1 = 1m
- h2 = 5m
- f = 1 GHz
- c = 3 × 10⁸ m/s

Find:

a. The Fresnel zone break point d_o.
b. The separation between two transmit antennas

Solution:

(a) $d_o = 4(10 - 1)\,(10 - 5)10^9/(3 \times 10^8 \text{ m/s})$

   $= 4 \times 9 \text{ m} \times 5 \text{ m} \times 10/3\text{m}$

   $= 600 \text{ m}$

(b) The above solution indicates that there exists a Fresnel zone break point at 600 m from the transmit antenna. Therefore, we need to install transmit antennas every $2d_o = 1200$ m in the above subway system.

3.7   Conclusions

- We have examined the Fresnel Zone Effects and various anomalies of RF propagation and have shown that there exists a free space propagation medium in multipath environments.
- We have presented a two ray model for outdoor deployment and have shown that these propagation models also exhibit equation of straight line within the Fresnel zone break point.
- We have also presented a two ray model for indoor deployment and a multi-ray propagation model
- For tunnels and subways and have shown that these propagation models also exhibit equation of straight line within the Fresnel zone break point.

References

1. David R. Smith, "Digital Transmission Systems", Van Nostrand Reinhold Co. ISBN: 0442009178, 1985
2. Theodore S. Rappaport, "Wireless Communications", Pearson Education, ISBN: 81-7808-648-4, 2002.
3. S. Faruque, "Cellular Mobile Systems Engineering", Artec House Inc., ISBN: 0-89006-518-7, 1996.
4. William C.Y. Lee, "Mobile Cellular Telecommunications Systems", McGraw- Hill Book Company, New York.

Chapter 4
Empirical Propagation Models

Objectives

- Review Empirical Propagation Models such as Lee Model, Okumura-Hata Model and Walfisch-Ikegami Model
- Show that Empirical Propagation Models also Exhibit Equation of Straight Line.
- Present radio frequency deployment guidelines

4.1 Empirical Propagation Models

The empirical models are based on extensive experimental data and statistical analysis which enable us to compute the received signal level in a given propagation medium. Many commercially available computer aided prediction tools are based on these models.

Among numerous propagation models, the following are the most significant ones, providing the foundation of today's land-mobile communication services [1]:

- Okumura-Hata Model
- Walfisch-Ikegami Model
- Lee Model

The usage and accuracy of these prediction models, however, depends chiefly on the propagation environment. For example, the standard Okumura-Hata model generally provides a good approximation in urban and suburban environments. On the other hand, the Walfisch-Ikegami model is applicable to dense-urban environments. This model is also useful for micro-cellular systems where antenna heights are generally lower than building heights, thus simulating an Urban Canyon environment.

The purpose of this section is to examine these models and show that all propagation models exhibits equation of straight lined having an intercept and a slope. We then classify the propagation environments in to four categories:

© Springer International Publishing Switzerland 2015
S. Faruque, *Radio Frequency Propagation Made Easy*, SpringerBriefs in Electrical and Computer Engineering, DOI 10.1007/978-3-319-11394-4_4

- Dense urban
- Urban
- Suburban
- Rural

Each propagation environment has a unique pathloss slope. Let's take a closer look:

4.2 Okumura-Hata Urban and Dense Urban Model Exhibits an Equation of Straight Line

The Hata model [2] is based on experimental data collected from various urban environments having approximately 16% high-rise buildings. The general path loss formula of the model is given by

$$L_p (dB) = C_o + C_1 + C_2 \log(f) - 13.82 \log(h_b) - a(h_m)$$
$$+ \left[44.9 - 6.55 \log(h_b) \right] \log(d) \qquad (4.1)$$

Where

L_p	path loss in dB
C_o	0 for Urban3 dB for Dense Urban
C_1	69.55 for 150 MHz$\leq$f$\leq$1000 MHz46.3 for 1500 MHz$\leq$f$\leq$2000 MHz
C_2	26.16 for 160 MHz$\leq$f$\leq$1000 MHz33.9 for 1600 MHz$\leq$f$\leq$2000 MHz
F	Frequency in MHz
h_b	Effective height of the base station in meters [30 m < hb < 30 m]
$a(h_m)$	$\{1.1 \log(F) - 0.7\} h_m - \{1.56 \log(F) - 0.8\}$—for Urban = 3.2[log(11.75h_m)]^2 − 4.97—for Dense Urban
h_m	Mobile antenna height [1 m < h_m < 10 m]
d	Distance between the base station and the mobile(km) [1 km < d < 20 km]

Equation (4.1) may be expressed conveniently as

$$L_p(dB) = L_o(dB) + \left[44.9 - 6.55 \ \log(h_b) \right] \log(d) \qquad (4.2)$$

or more conveniently as

$$L_p(dB) = L_o \ (dB) + \gamma 10 \log \ (d) \qquad (4.3)$$

The above equation exhibits an equation of a straight line, having an intercept L_o and a slope γ:

$$\text{Intercept}: \ \ L_o(dB) = C_0 + C_1 + C_2 \ \log(f) - 13.82 \log(h_b) - a(h_m) \qquad (4.4)$$

$$\text{Slope}: \ \ \gamma = \left[44.9 - 6.55 \log(h_b) \right] / 10 \qquad (4.5)$$

Fig. 4.1 a Shows two intercept points corresponding
to typical urban and dense
urban environments, having
identical pathloss slopes. **b**
Pathloss characteristics for
Okumura-Hata urban and
dense urban models

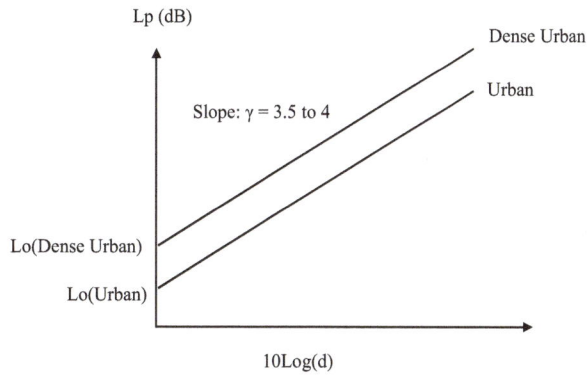

Note that there are two values for C_o:

- $C_o = 0$ dB for urban and
- $C_o = -3$ dB for dense urban

Also, there are two values for $a(h_m)$:

- $a(h_m) = \{1.1\log(F) - 0.7\}h_m - \{1.56\log(F) - 0.8\}$ for urban
- $a(h_m) = 3.2\left[\log\{11.75h_m\}\right]^2 - 4.97$ for dense urban
- $h_m =$ Mobile antenna height $[1\text{ m} < h_m < 10\text{ m}]$

Therefore, the intercept L_o has two values, one for urban and the other for dense
urban. On the other hand, the slope γ is the sane for both urban and dense urban.
It only depends on the base station antenna height (Fig. 4.1). Therefore, we write,

$$\text{For Urban}: \quad L_p(\text{dB}) = L_o(\text{Urban}) + \gamma 10\log(d) \tag{4.6}$$

$$\text{For Dense Urban}: \quad L_p(\text{dB}) = L_o(\text{Dense Urban}) + \gamma 10\log(d) \tag{4.7}$$

The slope γ is a function of the base station antenna height. Figure 4.2 shows that in
a typical urban and dense urban environment the attenuation slope varies between
3.5 and 4.

Fig. 4.2 Attenuation slope
as a function of base station
antenna height in a typical
urban and dense urban environment (due to Hata)

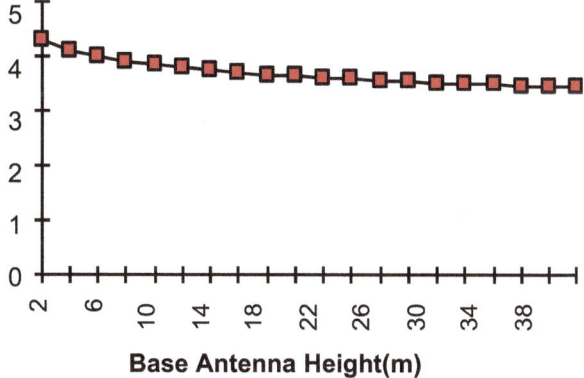

4.3   Okumura-Hata Suburban & Rural Model Exhibits an Equation of Straight Line

Hata suburban and rural models are based on the urban model with the following corrections:

$$L_p(\text{Suburban}) = L_p(\text{Urban}) - 2\left[\log(f/28)\right]^2 - 5.4 \qquad (4.8)$$

$$L_p(\text{rural}) = L_p(\text{Urban}) - 4.78\left[\log(f)\right]^2 + 18.33\log(f) - 40.94 \qquad (4.9)$$

Substituting for $L_0(\text{Urban})$ in the above equations, we get,

$$L_p(\text{Suburban}) = L_0(\text{Urban}) + \gamma 10\log(d) - 2\left[\log(f/28)\right]^2 - 5.4$$
$$= L_0(\text{Suburban}) + \gamma 10\log(d) \qquad (4.10)$$

$$L_p(\text{rural}) = L_0(\text{Urban}) + \gamma 10\log(d) - 4.78\left[\log(f)\right]^2 + 18.33\log(f) - 40.94$$
$$= L_0(\text{Rural}) + \gamma 10\log(d) \qquad (4.11)$$

The intercepts are given by,

$$L_0(\text{Suburban}) = L_0(\text{Urban}) - 2\left[\log(f/28)\right]^2 - 5.4 \qquad (4.12)$$

$$L_0(\text{Rural}) = L_0(\text{Urban}) - 4.78\left[\log(f)\right]^2 + 18.33\log(f) - 40.94 \qquad (4.13)$$

These suburban and rural models also follow the equation of a straight line, having different intercepts, where the pathloss slope remains the same. Figure 4.3 shows the pathloss characteristics for the suburban and the rural models.

Fig. 4.3 Pathloss characteristics for Okumura-Hata suburban and the rural models

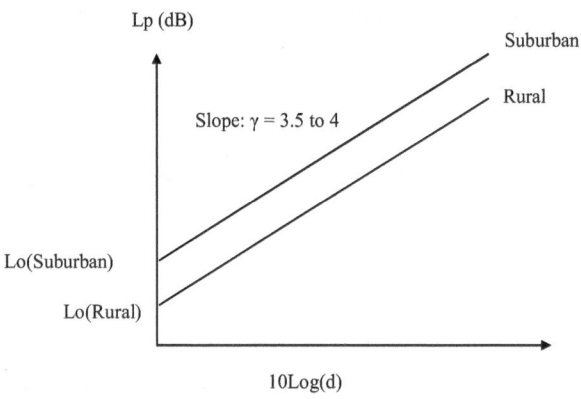

4.4 Walfisch-Ikegami Line of Sight (LOS) Model Exhibits an Equation of Straight Line

The Walfisch-Ikegami LOS model [3] is useful for dense urban environments. This model is based on several urban parameters such as building density, average building height, street widths etc. Antenna height is generally lower than the average building height, so that the signals are guided along the street, simulating an Urban Canyon type environment.

For Line Of Sight (LOS) propagation, the path loss formula is given by:

$$L_p(LOS) = 42.6 + 20\log(f) + 26\log(d) \qquad (4.14)$$

Where,

- f is the frequency in MHz
- d is the distance in km

The above equation can be described by means of the familiar "equation of straight line" as

$$L_p(LOS) = L_o + \gamma 10\log(d) \qquad (4.15)$$

Where L_o is the intercept and γ is the attenuation slope defined as

$$L_o = 42.6 + 20\log(f) \qquad (4.16)$$

$$\gamma = 2.6 \qquad (4.17)$$

Such a low attenuation slope in urban environments ($\gamma = 2.6$) is believed to be due to low antenna heights (below the roof top), generating wave-guide effects along the street. It follows that if a cell site is located at the intersection of a four way street, the contour of constant path loss would look like a diamond as shown in Fig. 4.4. Note that $\gamma = 2$ in free space.

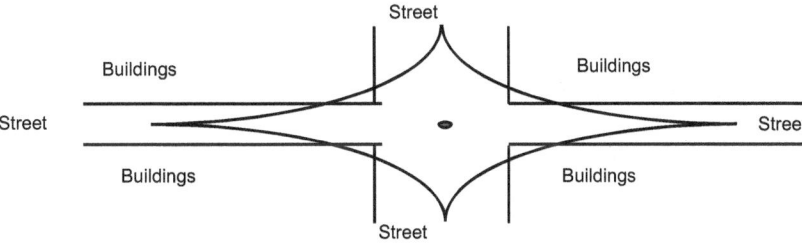

Fig. 4.4 Diamond shape coverage in dense urban canyon

4.4.1 Walfisch-Ikegami Non Line of Sight (NLOS) Model Exhibits Equation of Straight Line

For Non Line Of Sight (NLOS) propagation, the path loss formula is given by,

$$L_p(\text{NLOS}) = L_p(\text{Free Space}) + + L(\text{diff}) + L(\text{mult}) \tag{4.18}$$

Notice that the above equation also exhibits an equation of a straight line, because the free space pathloss exhibits an equation of a straight line and L(mult) and L(dif.) are constants. Let's take a closure look:

We define,

Lp (Free Space) $32.5 + 20 \log(f) + 20\log(d)$
f, d   Frequency and distance respectively.
L(diff.)   Roof-top diffraction loss
L(mult)   Multiple diffraction loss due to surrounding buildings

The rooftop diffraction loss is characterized as

$$L(\text{diff}.) = -16.9 - 10\log(\Delta W) + 10\log(f) + 20\log(\Delta h_m) + L(\emptyset) \tag{4.19}$$

The parameters in the above equation are defines as

ΔW Distance between the street mobile and the building
h_m Mobile antenna height
Δh_m $h_{roof} - hm$
$L(\emptyset)$ Loss due to elevation angle

The above parameters are constants after the antenna is installed. Therefore, L(diff.) is constant.

Multiple diffraction and scattering components are characterized by following equation:

$$L(\text{mult}) = k_o + k_a + k_d \cdot \log(d) + k_f \cdot \log(f) - 9\log(W) \tag{4.20}$$

Where

k_o $-18\log(1 + \Delta h_b)$
k_a $54 - 0.8(\Delta h_b) \ d \geq 0.5 \ km54 \ -0.8(\Delta h_b) \ d \leq 0.5 \ km$
k_d $18 - 16 \ (\Delta h_b / h_{roof})$
k_f $-4 + 0.7[(f/925) - 1]$ for suburban$- 4 + 1.5[(f/925) - 1]$ for urban
W Street width
h_b Base station antenna height
h_{roof} Average height of surrounding small buildings ($h_{roof} < h_b$)
Δh_b $h_b - h_{roof}$

The above parameters are also constants after the antenna is installed.

We assumed that the base station antenna height is lower than tall buildings but higher than small buildings. Combining Eq. (6.52), (6.53) and (6.54) we obtain

Fig. 4.5 Attenuation slope as a function of base station antenna height in a typical dense urban environment. (due to Walfisch-Ikegami)

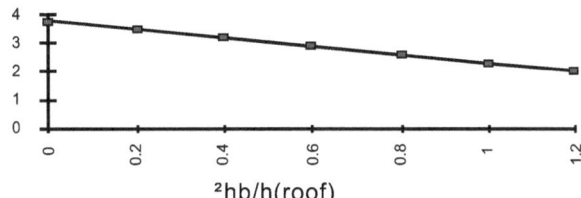

^{2}hb/h(roof)

$$L_p\left(NLOS\right) = L_o + \left(20 + k_d\right)\log(d)$$
$$= L_o + \gamma 10\log(d) \qquad (4.21)$$

The arbitrary constants are lumped together to obtain

$$L_o = 32.4 + \left(30 + k_f\right)\log(f) - 16.9 - 10\log(w) + 20\log\left(\Delta h_m\right)$$
$$+ L(\o) + k_o + k_a - 9\log(W)$$
$$\gamma = \left(20 + k_d\right)/10 \qquad (4.22)$$

Once again, the NLOS pathloss characteristics also exhibit an equation of straight line with L_o as the intercept and γ as the slope.

   The diffraction constant k_d depends on surrounding building heights, which vary from one urban environment to another, yielding a diffraction constant of a few meters to tens of meters. Typical attenuation slopes in these environments range from $\gamma = 2$ for $\Delta h_b/h_{roof} = 1.2$ to $\gamma = 3.8$ for $\Delta h_b/h_{roof} = 0$. This is shown in Fig. 4.5.

4.5   Lee Model

In order to accommodate terrestrial factors, C. Y. Lee has developed a simplified formula, given by [4]:

$$L_p = 129.45 + 38.4\,Log(d) - 20\,Log(H_b) \quad \text{for } 900\,MHz\ Cellular \qquad (4.23)$$

$$L_p = 135.45 + 38.4\,Log(d) - 20\,Log(H_b) \quad \text{for } 1900\,MHz\ Cellular\ (PCS)\ (4.24)$$

Where,

L_p Pathloss in dB
D   Distance in km
H_b Base station antenna height in meters
C   129.45 dB is the average loss in typical urban environment (f=900 MHz)
C   135.45 dB is the average loss in typical urban environment (f=1900 MHz)

The above equations can be conveniently written as follows:

$$L_p(900) = C1 + 38.4\,\text{Log}(d) \tag{4.25}$$

$$L_p(1900) = C2 + 38.4\,\text{Log}(d) \tag{4.26}$$

Once again, we see that Lee model also exhibits an equation of a straight line Where C1 and C2 are intercepts and γ is the slope:

C1 $129.45 - 20\,\text{Log}(H_b)$ (H_b = Base station antenna height)
C2 $135.45 - 20\,\text{Log}(H_b)$ (H_b = Base station antenna height)
γ 3.84 (slope)!!

4.6   Radio Frequency Deployment Guidelines

The empirical propagation models presented above, are based on extensive experimental data and statistical analysis which enable us to compute the received signal level in a given propagation medium. Yet, these propagation models, in practice, are fuzzy due to numerous rf barriers such as uneven terrain, buildings heights, hills, trees etc. building codes also vary from place to place. As a result, the accuracy of these prediction models depends on the frequency, antenna height and propagation environment. For example, the standard Okumura-Hata model generally provides a good approximation in urban and suburban environments. On the other hand, the Walfisch-Ikegami model is applicable to dense-urban environments. This model is also useful for micro-cellular systems where antenna heights are generally lower than building heights, thus simulating an Urban Canyon environment. Lee model can also be used for 900 MHz macro cell and 1900 MHz PCS system. We also note that all propagation models exhibits free space pathloss within the Fresnel zone break point. We then classified the propagation environments in to four categories: Dense urban, Urban, Suburban and Rural.

Each propagation environment has a unique pathloss slope. Table below provides a guideline to use these propagation models in various propagation environments (Table 4.1).

4.7   Conclusions

- We have presented a general overview of various empirical prediction models and have shown that these propagation models also exhibit equation of straight line within the Fresnel zone break point.
- Although these predictions and measurement techniques are the foundation of today's cellular services, they suffer from inaccuracies due to user defined clutter factors. These clutter factors arise due to numerous RF barriers which vary

Table 4.1 RF deployment guidelines

Propagation environments	Typical pathloss slope (γ)	Propagation models
Dense Urban	4	Walfisch-Ikegami Model
High-rise Buildings "canyon" channel propagation		
Antennas above the roof-top, causing multiple diffractions		
Antennas below the roof-top., causing multiple reflections		
Urban		Okumura-Hata Model
Mixture of various building heights and open areas		
Suburban	3	Okumura-Hata Model
Residential areas		
Open fields		
Rural	2.5	Okumura-Hata Model
Farm areas		
Highways		
Free space	2	Cell radii within the Fresnel zone break point
Outer space		
Terrestrial environments: Distance within the Fresnel zone break point. All propagation environments. Depends on frequency and antenna height		

from place to place. It is practically impossible to accommodate all these factors accurately. Cell site location is also a challenging engineering task because of regulations and restrictions imposed on some locations. Therefore cell sites have to be relocated from the predicted location, requiring best judgment of RF engineers. Thus we came to the conclusion that propagation prediction is a combination of science, engineering and art. An experienced RF engineer, willing to compromise between theory and practice, is expected to accomplish the most.

References

1. S. Faruque, "Cellular Mobile Systems Engineering", Artec House Inc., ISBN: 0–89006-518–7, 1996.
2. Hata. M, " Empirical formula for propagation loss in land mobile radio services", IEEE Trans. on Vehicular Technology, VT-29, pp 317–326, 1980.
3. Walfisch, J et. al' "A theoretical model of UHF Propagation in Urban Environments", IEEE Trans. on Antenna and Propagation, AP-38, pp1788–1796, 1988.
4. William C.Y. Lee, "Mobile Cellular Telecommunications Systems", McGraw- Hill Book Company, New York.

Chapter 5
Statistical Analysis in RF Engineering

5.1 Why Statistical Analysis?

Statistics is the s study of the collection, organization, analysis, interpretation and presentation of data [1–8]. For Radio Frequency (RF) engineering it involves live air data collection as a function of distance. Since multipath propagation is fuzzy owing to numerous RF barriers, uneven terrain, hills, trees, buildings etc., there is a large variation of received signal Level (RSL) at the receiver [9–12].

Figure 5.1 shows an example to illustrate this scenario. Here, the received signal level (RSL) is measured in dBm and the distance is measured in km. Notice that the signal strength decays logarithmically as a function of distance where the distance is plotted in the linear scale.

Also notice that we have plotted a solid line, which is known as the regression line or the best fit. The regression line has a special significance since 50% data are above the regression line and 50% data are below the regression line. Furthermore, at a given distance, the distribution of data has a shape known as Gaussian (Bell shaped). We shall discuss these points along with its attributes latter in this chapter.

5.2 Regression Analysis

Regression analysis is a statistical process for estimating the relationships among variables [13–17]. For RF engineering, it includes distance as the independent variable and RSL as the dependent variable. Here, we are interested in the dependence of RSL on distance d; we generally refer to it as the regression curve of RSL on d.

In the experiment we select n values of distances d_1, d_2, ... , d_n and monitor the corresponding RSL value yielding paired samples as follows:

$$(d_1, RSL_1), (d_2, RSL_2), ...(d_n, RSL_n) \qquad (5.1)$$

© Springer International Publishing Switzerland 2015
S. Faruque, *Radio Frequency Propagation Made Easy,* SpringerBriefs in Electrical and Computer Engineering, DOI 10.1007/978-3-319-11394-4_5

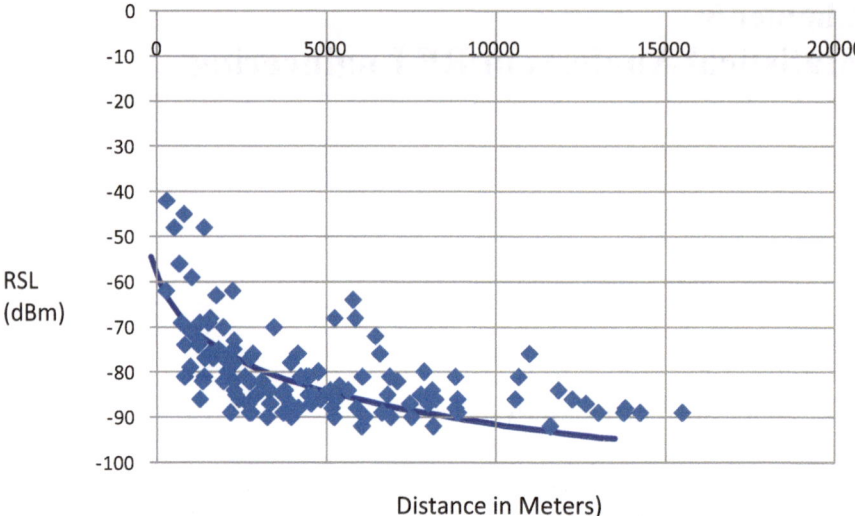

Distance in Meters)

Fig. 5.1 Received signal strength as a function of distance in semi-logarithmic scale

In regression analysis the mean of RSL is a linear function of the distance. With Fresnel zone point this is given by,.

$$\overline{RSL} = RSL_o - \gamma d/d_o \tag{5.2}$$

Where RSL_o is the Intercept, γx is the slope and d_o is the Fresnel zone break point. In logarithmic scale, it may be expressed as,

$$\overline{RSL}(dB) = RSL_o(dB) - 10\gamma \log(d/d_o) \tag{5.3}$$

The above equation is plotted in Fig. 6.11 where RSL_o is the received signal level (intercept) at the break point. d_o and γ is the slope after d_o. The mean $\overline{RSL}$ is related to the random variable $(RSL)_i$ as (Fig. 5.2)

$$(\Delta RSL)_i = (RSL)_i - \overline{RSL}$$
$$= (RSL)_i - RSL_o + 10\ \gamma \log[(d_i/d_o)] \tag{5.4}$$

Where $(\Delta RSL)_i$ is the deviation of the random variable $(RSL)_i$ from the mean $\overline{RSL}$. In order to fit a straight line, it is desirable to reduce $(\Delta RSL)_i$. In other words, the sum of square of this difference must be equal to zero. Thus we write,

$$\sum_{i=1}^{n} (\Delta RSL_i)^2 = 0 \tag{5.5}$$

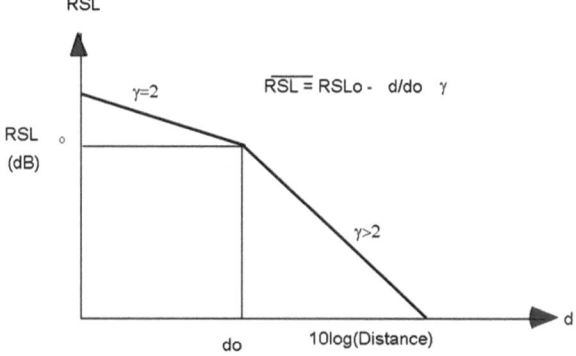

Fig. 5.2 Regression curve in a given propagation environment

Where

$$\sum_{i=1}^{n}\left(\Delta RSL_i\right)^2 = \sum_{i=1}^{n}\left[\left(RSL\right)_i - RSL_o + 10\gamma\log\left(d_i/d_o\right)\right]^2 \tag{5.6}$$

Then

$$\frac{\partial \sum \left(\Delta RSL_i\right)^2}{\partial RSL_o} = 0 \text{ and } \frac{\partial \sum \left(\Delta RSL_i\right)^2}{\partial (d_i/d_o)} = 0 \tag{5.7}$$

or

$$-2\sum_{i=1}^{n}\left[\left(RSL\right)_i - RSL_o + 10\gamma\log\left(d_i/d_o\right)\right] = 0 \tag{5.8}$$

$$-2\sum_{i=1}^{n}\left[\left(d_o/d_i\right)\left\{\left(RSL\right)_i - RSL_o + 10\gamma\log\left(d_i/d_o\right)\right\}\right] = 0 \tag{5.9}$$

This reduces to

$$\sum_{i=1}^{n}\left(RSL\right)_i = n.RSL_o - 10\gamma\sum_{i=1}^{n}\log\left(d_i/d_o\right) \tag{5.10}$$

$$\sum_{i=1}^{n}\left(d_o/d_i\right)\cdot\left(RSL\right)_i = RSL_o\sum_{i=1}^{n}\left(d_o/d_i\right) - 10\gamma\left[\sum_{i=1}^{n}\left(d_o/d_i\right)\cdot\log\left(d_i/d_o\right)\right] \tag{5.11}$$

Solving the above equations, we obtain RSL_o and γ in a given propagation environment [18]. The break point is given by $d_o = 4h_1h_2/\lambda$, where h_1 and h_2 are the transmit and receive antenna heights respectively. The above method is known as method of

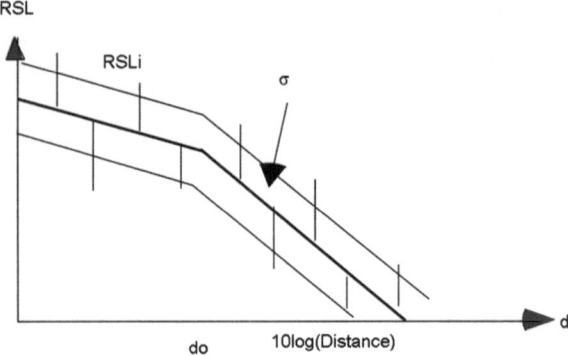

Fig. 5.3 RSL as a function of distance. Solid lines are due to regression fit. do is the break point and s is the standard deviation

least squares, which can be used to fit a straight line to a given set of paired data (d_i, RSL_i).

5.3   Prediction of Random Data with Confidence

In statistical analysis we often induce a generalization from a set of random variables. For example, the regression analysis presented in the previous section is based on a set of paired data (d_i, RSL_i), where RSL_i is the random variable. It provides a point estimation from a random sample, which enables us to estimate the received signal level at a given distance, with reasonable accuracy within a certain range of standard deviation. This range is known as Confidence Interval. Once we have a confidence interval, we can reasonably assure ourselves that the mean of RSL will exist within this interval with a certain probability. This probability is known as Confidence Level, which varies between 0 and 1 (0–100%).

Figure 5.3 shows the dependence of RSL as a function of distance, along with the regression line. Since we cannot expect to coincide the regression line with the actual straight line, our interest is to determine an interval within which the regression line may exist with a high probability.

Now we consider a set of random variables RSL_i having n sample values where $i = 1,2,...$ n. The distribution or the density of such a set of random numbers is generally approximated by a continuous curve known as Normal Distribution. The equation that describes a normal distribution is given by [9–10]

$$f(RSL) = \frac{1}{\sigma\sqrt{2\pi}}\exp\left(-0.5\left[\left(\frac{RSL - \overline{RSL}}{\sigma}\right)\right]^2\right) \qquad (5.12)$$

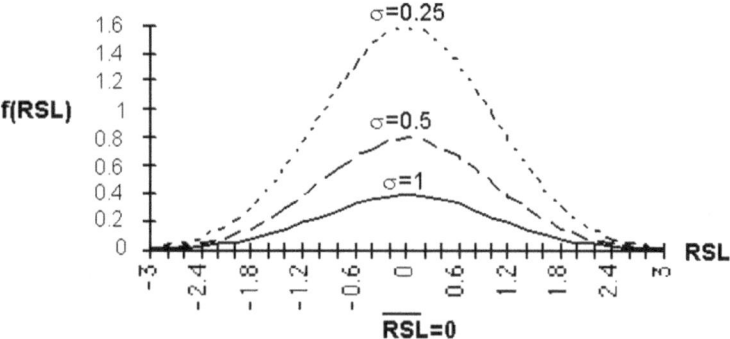

Fig. 5.4 Normal distribution with zero mean (RSL=0) and variable standard deviation

Where the mean is given by

$$\overline{RSL} = \frac{RSL_1 + RSL_2 + \ldots + RSL_n}{n} \tag{5.13}$$

And the variance is given by

$$\sigma^2 = \frac{(RSL_1 - \overline{RSL})^2 + (RSL_2 - \overline{RSL})^2 + \ldots + (RSL_n - \overline{RSL})^2}{n-1} \tag{5.14}$$

σ being the Standard Deviation.

The curve of Fig. 5.4 is also known as the Gaussian Distribution or a bell shaped curve which is symmetric with respect to the mean whose peak at $\overline{RSL} = 0$ increases as σ decreases.

Figure 5.5 shows the distribution curve for $\overline{RSL} \neq 0$. We notice that for a positive mean, the curve has the same shape but is shifted to the right, and for a negative mean, it is shifted to the left. This illustrates the fact that the variance is the average dispersion from the mean.

The probability density function of Eq. (6.70) is generally obtained from the standard table called Standard Normal Distribution, or by means of a curve called Cumulative Distribution Function as shown in Fig. 5.6. Both are based on the following probability distribution function:

$$F(z) = \frac{1}{\sigma\sqrt{2\pi}} \int_{-\infty}^{z} \exp\left(-0.5\left[\left(\frac{RSL - \overline{RSL}}{\sigma}\right)\right]^2 d(RSL)\right) \tag{5.15}$$

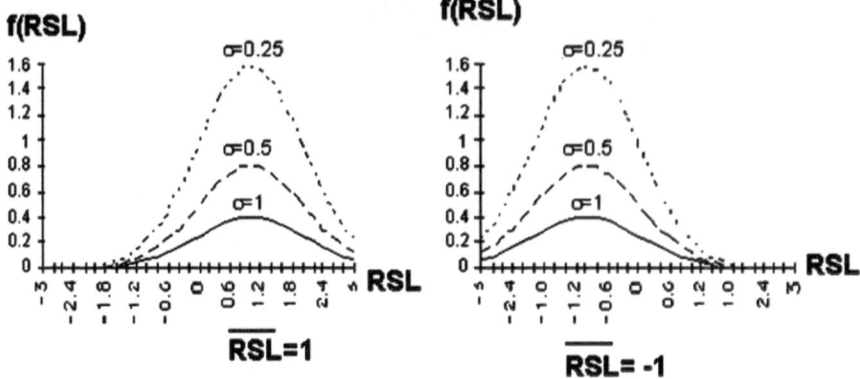

Fig. 5.5 Normal distribution with RSL=1 and RSL=−1, s=variable

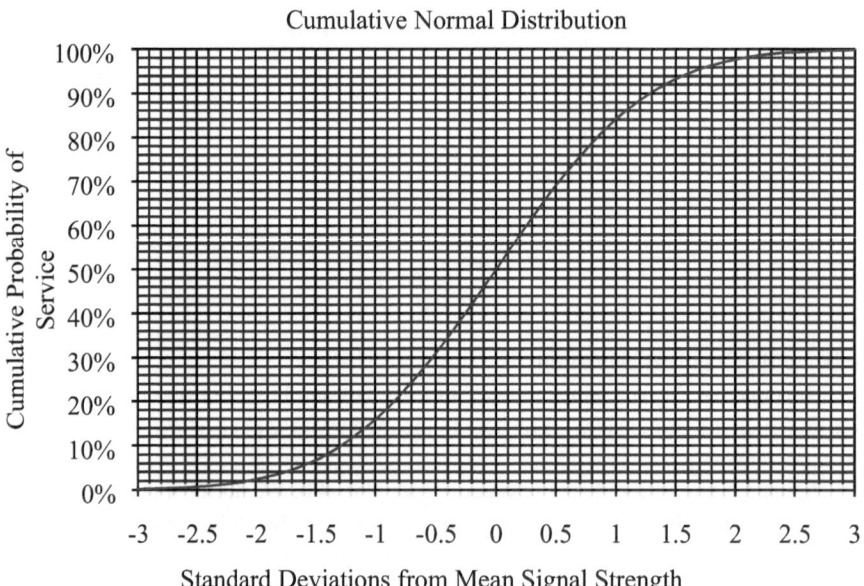

Fig. 5.6 Cumulative probability distribution as a function of normalized standard deviation(z)

With $\overline{RSL} = 0$ and $\sigma = 1$. Then the random variable (RSL) can be estimated from the following normalized standard deviation z where σ is the measured standard deviation.

$$z = \left(\frac{RSL - \overline{RSL}}{\sigma} \right)$$ (5.16)

or

$$RSL = \sigma z + \overline{RSL} \qquad\qquad (5.17)$$

5.3.1   Problem 5.1

Given:

- Desired received signal RSL$=-80$ dBm
- Measured standard deviation $\sigma=8$ dB,
- Required confidence level is 80%

Determine the minimum received signal level (RSL) to satisfy the above requirements.

5.3.2   Solution 5.1

For 80% confidence level, using the curve in Fig. 5.6 we have,

$$z(0.8) = 0.842.$$

Therefore, the minimum RSL can be computed as,

$$
\begin{aligned}
RSL &= \sigma z + \overline{RSL} \\
&= (8 \times 0.842) - 80\,\text{dBm} \\
&= -73.26\,\text{dBm}.
\end{aligned}
$$

This is the signal strength at the cell edge, which ensures that 80% of the data will fall within the interval $-\sigma$ and $-\sigma$, i.e. within $+8$ dB. This interval is called the confidence interval and the probability (80%) is called the confidence level.

5.4   Drive Test, Data Collection and Statistical Analysis

For cellular communications, RF engineering involves live air data collection as a function of distance and statistical analysis of the data. Today, a large number of pc based data collection tools are commercially available. These data collection tools have the capability to import measurement data and generate statistical outputs such as mean error, standard deviation, max., min., etc.

Figure 5.7 shows the basic concept of RF data collection technique. It is PC based and uses a cellular radio, MS Excel, a GIS(Geographic Information Services) software, and a GPS (Global Positioning System) receiver. The GPS receiver is

Fig. 5.7 Drive test and data collection technique. The received signal level (RSL) is measured as a function of distance

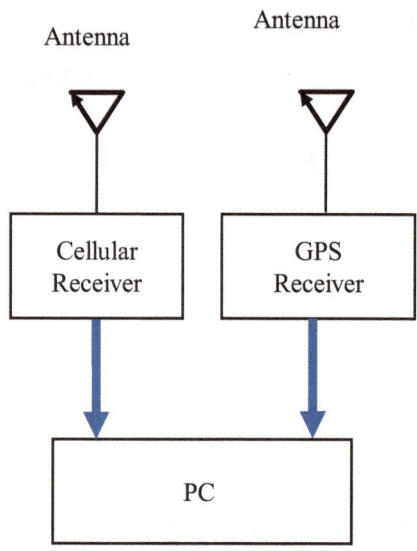

used to collect the coordinates (longitude & latitude) of each sampling points. The outcome is a pair of long/lat corresponding to each RSL value. Since the cell site location is fixed and has a unique long/lat value, the distance of each sampling point with respect to the cell site is readily available as an output.

Table 5.1 shows an output file, which was obtained by means of drive test. Notice that the Received Signal Level (RSL) is measured in dBm as a function of distance, where the distance is in meters. We can now perform statistical analysis to find the following parameters:

- Mean
- Standard deviation
- Minimum RSL value
- Maximum RSL value and
- The regression line

The above statistical parameters were calculated in Excel and also presented in Table 5.1 at the end.

Note that the above data was collected from a cell site in a typical urban environment. Our analysis indicates that the cell site exhibits following performance characteristics:

Mean RSL $=-79.8$ dBm
Standard Dev. $=10.08$ dB
Max. RSL $=-42$ dBm
Mi. RSL $=-92$ dBm

These values are typical in urban environment and the cell site is healthy. The 10 dB standard deviation has a special significance in designing reliable cell sites, which we shall see later through a practical design.

Table 5.1 Output file obtained from drive test.

Distance_(m)	RSL(dBm)_900 MHz	
269	−62	
1265	−86	
667	−56	
2437	−86	
1658	−77	
2289	−78	
13041	−89	
7791	−85	
5268	−86	
2080	−80	
1553	−76	
4398	−81	
4760	−80	
976	−79	
1579	−76	
2611	−81	
4463	−85	
8818	−81	
6574	−76	
8118	−84	
1403	−81	
6860	−81	
3464	−70	
2232	−62	
6798	−85	
6636	−89	
2745	−82	
3776	−84	
2852	−76	
8821	−88	
15481	−89	
11864	−84	
1961	−70	
5220	−90	
12264	−86	
1360	−82	
6891	−90	
2773	−82	
2175	−89	

Table 5.1 (continued)

Distance_(m)	RSL(dBm)_900 MHz	
2756	−77	
1522	−69	
5230	−68	
3719	−85	
1260	−73	
293	−42	
826	−81	
1968	−82	
512	−48	
1279	−69	
1005	−71	
2799	−89	
3312	−84	
2287	−84	
1763	−63	
5640	−84	
3357	−87	
7877	−80	
8153	−92	
6049	−92	
731	−69	
1413	−77	
3948	−78	
1401	−48	
14261	−89	
3266	−90	
1584	−68	
8003	−87	
1029	−59	
6447	−72	
3936	−87	
6066	−81	
5780	−64	
12674	−87	
13832	−88	
3745	−89	
13775	−89	
803	−45	
5156	−88	
6135	−90	
8218	−86	

Table 5.1 (continued)

Distance_(m)	RSL(dBm)_900 MHz	
1222	−74	
11008	−76	
10682	−81	
2761	−82	
4479	−81	
2268	−82	
4925	−85	
7096	−82	
3116	−82	
4536	−87	
5385	−83	
4236	−81	
11618	−92	
4160	−76	
4813	−86	
4627	−85	
10593	−86	
8895	−89	
5849	−68	
7460	−87	
5878	−88	
2714	−87	
2945	−85	
822	−74	
2740	−89	
1827	−75	
1984	−77	
2273	−75	
2285	−73	
5103	−84	
8883	−86	
2339	−85	
4167	−88	
3967	−90	
7505	−90	
Mean=	−79.87826087	dBm
Stdev=	10.07963713	dB
Max=	−42	dBm
Min=	−92	dBm

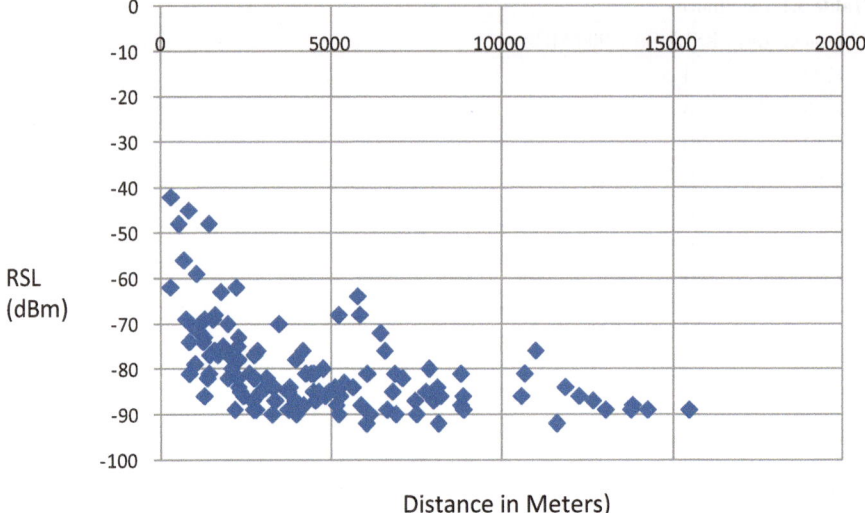

Fig. 5.8 The received signal Level (RSL) as a function of distance in a semi-logarithmic scale. RSL is measured in dBm and the distance is measured in meters

We now turn our attention to Fig. 5.8, where RSL is plotted in dBm and the distance is plotted in the linear scale. Notice that the signal strength decays logarithmically as a function of distance where the distance is plotted in the linear scale. The rate of decay depends on the propagation environment.

Next, we would like to see the regression line, which is only one click away. This is an excel feature and is shown in Fig. 5.9. The solid line in this figure is the regression line, which is also known as the best fit. As known in this art, 50 % points are above the regression line and 50 % points are below the regression line. This means that if we define the cell radius by means of the regression line, 50 % RSL values will be within the cell and 50 % RSL values will be outside the cell. That is, the cell is 50 % reliable. Let's examine this by means of the following problem:

5.4.1   Problem 5.2: Cell Design With 50 % Confidence Level

You are assigned to design a cell site to deliver at least −80 dBm to at least 50 % of the locations in an area. Measurements you've made have the following parameters:

- Mean RSL = −80 dBm
- Standard deviation = 10 dB

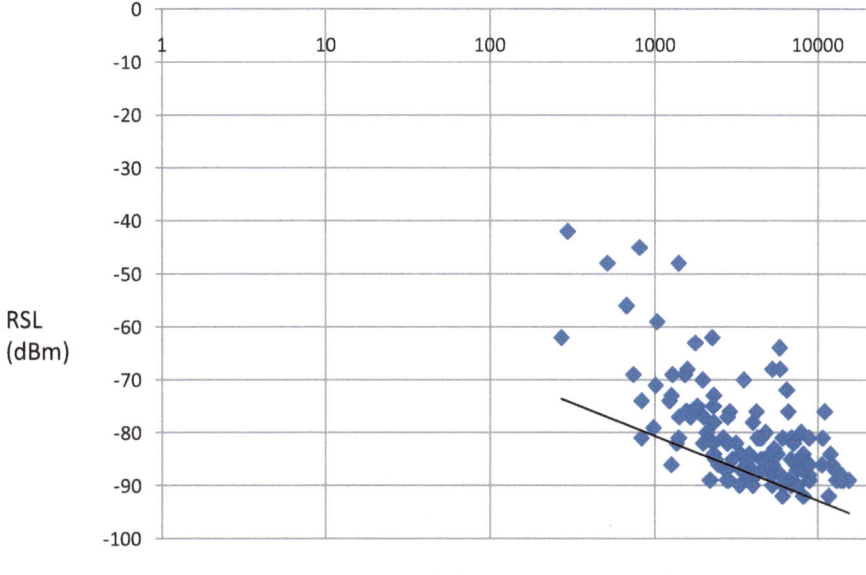

Fig. 5.9 The received signal Level (RSL) as a function of distance in Log-Log scale. RSL is measured in dBm and the distance is measured in 10Log (Distance)

5.4.2   Solution 5.2:

For 50% confidence level, using the curve in Fig. 5.6 we have,

$$z(0.5) = 0$$

Therefore, with − 10 dB standard deviation, the minimum RSL can be computed as,

$$RSL = \sigma z + \overline{RSL}$$
$$= (10 \times 0) - 80\, dBm$$
$$= -80\, dBm.$$

This is the signal strength at the cell edge, which ensures that 50% of the data will fall within the interval $-\sigma$ and $+\sigma$, i.e. within ± 10 dB. This interval is called the confidence interval and the probability (50%) is called the confidence level.

5.4.3   Problem 5.3: Cell Design With 90% Confidence Level

You are assigned to design a cell site to deliver at least − 80 dBm to at least 90% of the locations in an area. Measurements you've made have the following parameters:

- Mean RSL$=-80$ dBm
- Standard deviation$=10$ dB

5.4.4   Solution 5.3:

For 90 % confidence level, using the curve in Fig. 5.6 we have,

$$z(0.9) = 1.3$$

Therefore, with -10 dB standard deviation, the minimum RSL can be computed as,

$$\text{RSL} = \sigma z + \overline{RSL}$$
$$= (10 \times 1.3) - 80\,\text{dBm}$$
$$= -67\,\text{dBm}.$$

This is the signal strength at the cell edge, which is stronger than the one in the previous problem. It ensures that 90 % of the data will fall within the interval $-\sigma$ and $+\sigma$, i.e. within ± 10 dB. This interval is called the confidence interval and the probability (90 %) is called the confidence level.

5.5   A PC Based RF Planning Tool: A Student Project

5.5.1   Background

Today, numerous computer-aided RF design tools are available for planning and designing the cellular system. These tools generally begin with empirical propagation models such as Okumura-Hata, Walfisch-Ikegami model, where the geographic information. is already built-in. Yet, most prediction tools still require GIS tool to run the prediction. In addition, drive test data is used for model tuning, thus defeating the original purpose. Moreover, these tools require user defined clutter factors, which are subjective; as a result an error is inevitably present in these tools. Last but not least, these tools are complex and expensive; users require specialized training to use them effectively.

The RF planning tool presented here differs from the others in that it is very inexpensive and simple to develop as a student project. It is PC based and uses Empirical propagation models, MS Excel, a GIS(Geographic Information Services) software, and a GPS (Global Positioning System) receiver. The GPS receiver will be used to collect the coordinates of each cell site (Lat/Long). Once the lat/long of antenna locations are known, the cell radii can be computed in Excel by using any of the existing empirical models such as Okumura-Hata or Walfisch-Ikegami.

Table 5.2 Empirical Propagation Models and their Usage

Environmental Zone	Commonly Used Models
Dense Urban	Walfisch-Ilkegami
Building "canyon" channel propagation	
Antennas above buildings (macro-cell) casuemultiple diffractions over buildings	Okumura-Hata
Antennas below buildings (micro-cell) causediffractions around and reflections on buildings	
Urban	Walfisch-Ilkegami
Mixture of various building heights and open areas	Okumura-Hata
Suburban	
Business and resential areas, open areas, woods	Okumura-Hata
Rural	
Large open areas	
Multiple diffractions over obstacles	Okumura-Hata

Finally, the MapInfo software will be used to import lat/long and the corresponding radius to display the RF coverage plot along with the geographic information such as roads, highways, water, park, buildings etc.

We believe that this project will be a valuable learning experience for the students. They will gain hands-on experience in project management, time management, teamwork and documentation.

5.5.2   Choice of Propagation Model

There are several empirical propagation models available to design cellular systems. These prediction models are based on extensive experimental data and statistical analysis, which enable us to compute the received signal level in a given propagation medium. Many commercially available computer aided prediction tools are based on these models. The usage and accuracy of these prediction models depend on the propagation environment. For example, the Okumura-Hata model generally provides a good approximation in urban and suburban environments, where the antenna is placed on the roof of the tallest building [19]. On the other hand, the Walfisch-Ikegami model can be applied to dense-urban and urban environments, where the antenna height is below the rooftop [20]. A brief description and usage of these models is presented in Table 5.2.

5.5.3   Description of the CAD Project

The RF planning tool presented below (Fig. 5.10) has been developed as a student project. It uses:

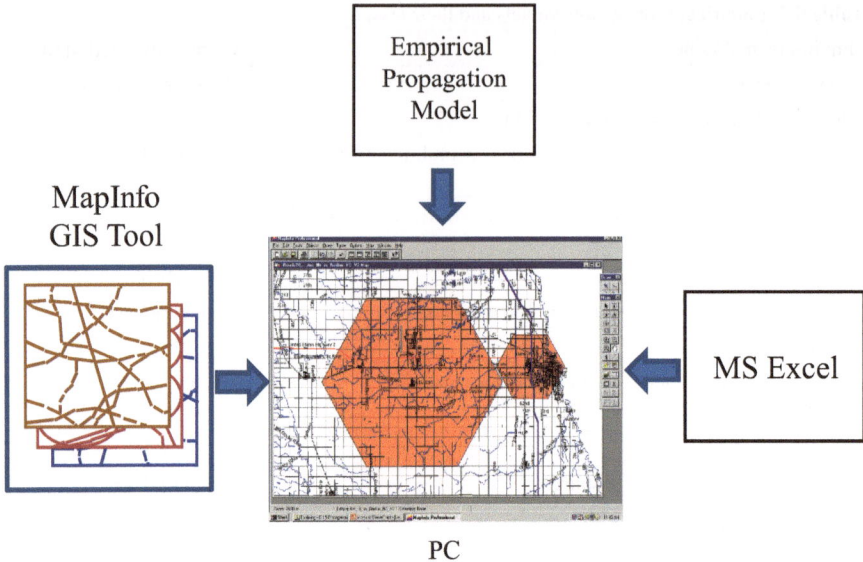

Fig. 5.10 Illustration of the PC based RF planning tool

- A GIS software such as MapInfo
- Empirical propagation models,
- MS Excel and
- A PC.

The GIS (Geographic Information Software) tool imports the cell radius. Empirical propagation models, such as Okumura-Hata, Walfisch-Ikegami, Lee etc. provide pathloss characteristics in a given propagation environment. The MS Excel compute cell radius from a given propagation model. It also tabulates Long/Lat, antenna height, Cell radii etc. Upon receiving the lat/long, the PC-based RF prediction tool displays the RF coverage along with the geographic information.

5.5.4 Step-By-Step Design Process

Step-1: Collect Latitude and Longitude (Lat/Long) using GPS receiver and determine base station antenna height. This is an outdoor activity involving:

- Site visits
- Identifying base station location
- Recording Lat/Long using GPS receiver
- Record antenna height

Step-2: Choose a propagation model from Table 5.2 and compute the required path loss L_p and the corresponding received signal:

- Path Loss: $L_p = 134.2$ dB (from the chosen model)
- Received Signal = EIRP − Lp (will vary from cell site to cell site)

Step-3: Tabulate design parameters. This step involves collecting and tabulating various design parameters listed below:

- Frequency
- Propagation model
- Base station power
- Antenna gain
- Signal strength at the cell edge
- Mobile antenna height
- Pathloss Lp (from the propagation model)

Step-4: Compute cell radius as a function of antenna height. As an example, consider the Hata pathloss model:

- $L_p (dB) = C_1 + C_2 \log(f) - 13.82 \log(h_b) - a(h_m) + [44 \cdot 9 - 6 \cdot 55 \log(h_b) \cdot] \log(d) + C_o$

- Solving for cell radius (d) we obtain

- Cell Radius: $d = 10^{\dfrac{L_p (dB) - C_o [C_1 + C_2 \log(f) - 13.82 \log(h_b) - a(h_m)]}{[44.9 - 6.55 \log(h_b).]}}$

Step-5: Repeat Step-4 for each location (Lat/Long) and generate an Excel spread sheet. This spread sheet produces cell radii as a function of antenna height for each location as follows:

Longitude	Latitude	Antenna Height (m)	Cell Radius (km)	Propagation environment
−97.0958	47.9202	15	6.069	Urban
−97.376	47.8978	30	15.961	Suburban

Step-6: Import Lat/Long and the corresponding Cell Radius from the above table into the GIS tool (e.g. MapInfo) using the specific set of commands, supported by the GIS software. It is assumed that the GIS software has already been installed into the PC and it is up and running. At this point, each lat/long and the corresponding cell radius will be imported instantly. The outcome is a composite RF coverage plot superimposed on the map, as shown in Fig. 5.11.

5.6 Conclusions

- Reviewed Statistical Analysis and showed that it is an important exercise to design and implement cellular base stations with reliability.

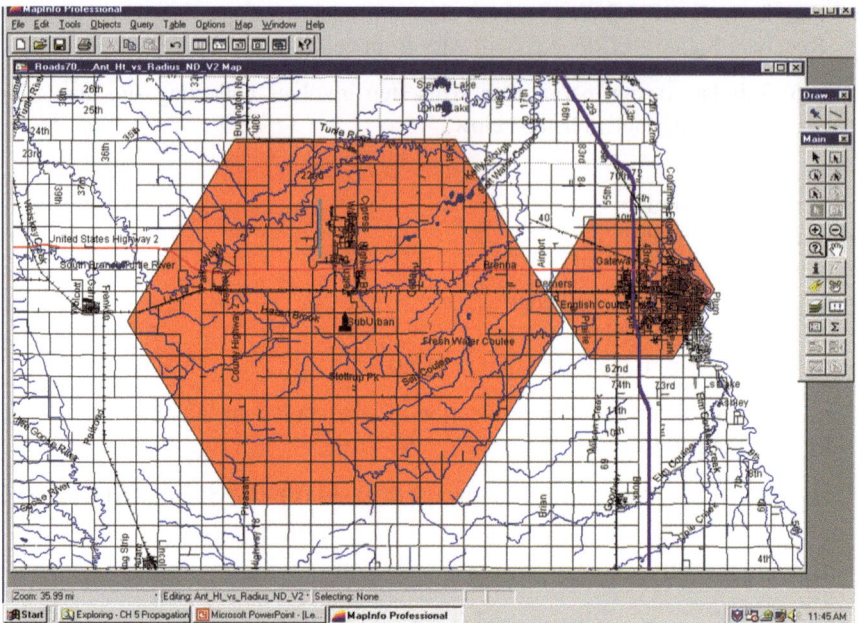

Fig. 5.11 Composite RF coverage plot on PC

- Presented regression analysis and showed that random data such as Received Signal Level (RSL) can be predicted with confidence.
- Drive Test, Live air data collection & Data analysis techniques were presented.
- A computer aided prediction technique was presented as a student project.

References

1. Irwin Miller and John E. Freund, "Probability and Statistics for Engineers", Prentice-Hal, Inc, 1977.
2. David A. Freedman, Statistical Models: Theory and Practice, Cambridge University Press (2005)
3. Mogull, Robert G. (2004). Second-Semester Applied Statistics. Kendall/Hunt Publishing Company. p. 59. ISBN 0-7575-1181-3.
4. Yule, G. Udny (1897). "On the Theory of Correlation". Journal of the Royal Statistical Society (Blackwell Publishing) 60 (4): 812–54. doi:10.2307/2979746. JSTOR 2979746.
5. Ronald A. Fisher (1954). Statistical Methods for Research Workers (Twelfth ed.). Edinburgh: Oliver and Boyd. ISBN 0-05-002170-2.
6. Steel, R.G.D, and Torrie, J. H., Principles and Procedures of Statistics with Special Reference to the Biological Sciences., McGraw Hill, 1960, page 288.
7. Chiang, C.L, (2003) Statistical methods of analysis, World Scientific. ISBN 981-238-310-7 - page 274 section 9.7.4 "interpolation vs extrapolation"
8. Good, P. I.; Hardin, J. W. (2009). Common Errors in Statistics (And How to Avoid Them) (3rd ed.). Hoboken, New Jersey: Wiley. p. 211. ISBN 978-0-470-45798-6.

9. IS-54, "Dual-Mode Mobile Station – Base Station Compatibility Standard," Electronic Industries Association Engineering Department, PN -2216, Dec. 1989.
10. William C.Y. Lee, "Mobile Cellular Telecommunications Systems", McGraw- Hill Book Company, New York.
11. L.B.Milstein et.al., "On the Feasibility of a CDMA Overlay for Personal Communications Network", IEEE Journal on Selected Areas in Communications, vol. 10, No. 4, May,1992, pp. 666–668.
12. H.H. Xia et. al., "Radio Propagation Characteristics for Line Of Sight Micro cellular and Personal Communications", "IEEE Transactions on Antennas and Propagation", Vol. 41, No. 10, October, 1993, pp. 1439–1447.
13. Armstrong, J. Scott (2012). "Illusions in Regression Analysis". International Journal of Forecasting (forthcoming) 28 (3): 689. doi:10.1016/j.ijforecast.2012.02.001.
14. R. Dennis Cook; Sanford Weisberg Criticism and Influence Analysis in Regression, Sociological Methodology, Vol. 13. (1982), pp. 313–361
15. Fisher, R.A. (1922). "The goodness of fit of regression formulae, and the distribution of regression coefficients". Journal of the Royal Statistical Society (Blackwell Publishing) 85 (4): 597–612. doi:10.2307/2341124. JSTOR 2341124.
16. Rodney Ramcharan. Regressions: Why Are Economists Obsessed with Them? March 2006. Accessed 2011-12-03.
17. Tofallis, C. (2009). "Least Squares Percentage Regression". Journal of Modern Applied Statistical Methods 7: 526–534. doi:10.2139/ssrn.1406472.
18. Saleh Faruque, "Cellular Mobile Systems Engineering", Artec House Inc., ISBN: 0-89006-518-7, 1996.
19. Hata. M, "Empirical formula for propagation loss in land mobile radio services", IEEE Trans. on Vehicular Technology, VT-29, pp 317–326, 1980.
20. Walfisch, J et. al' "A theoretical model of UHF Propagation in Urban Environments",

Chapter 6
Radio Frequency Coverage: The Cell

6.1 Introduction

The FCC (Federal Communications Commission) provides licenses to operate cellular communication systems over a given band of frequencies. These bands of frequencies are finite and have to be reused to provide services to other geographic areas. Since a number of different technologies are available [1, 2], the reuse of frequencies must also differ. Therefore, techniques are needed to support these competing technologies so that they can coexist and offer services to a wide geographic area without interfering each other. In order to provide a comprehensive overview, this chapter begins with the description of cell geometry followed by the concept of cell reuse with the evaluation of carrier to interference ratio (C/I). The classical cell reuse plan [3] is described next, with examples of various frequency plans related to OMNI and Sectorization schemes. The associated channel capacity and C/I performances are also evaluated.

6.2 The Concept of Cell

A cell, also known as the Radio Frequency (RF) footprint, is the basic building block in wireless communication. There are two types of cells:

- OMNI Cell and
- Sectorized cell

OMNI means all directions and uses OMNI directional antenna at the center of the cell as shown in Fig. 6.1a. Sectorized cells use directional antennas. Typically, there are three sectors, 120° per sector. Each sector uses a directional antenna as shown in Fig. 6.1b. Also shown in Fig. 6.1c is an alternate version of a sectorized cell known as tri-cellular plan. We will further discuss these cell structures later in this chapter.

Each OMNI and sectorized cell has a base station, where the transmitter is located at the center of the cell. The shape and size of the cell depends on transmit power,

© Springer International Publishing Switzerland 2015
S. Faruque, *Radio Frequency Propagation Made Easy,* SpringerBriefs in Electrical and Computer Engineering, DOI 10.1007/978-3-319-11394-4_6

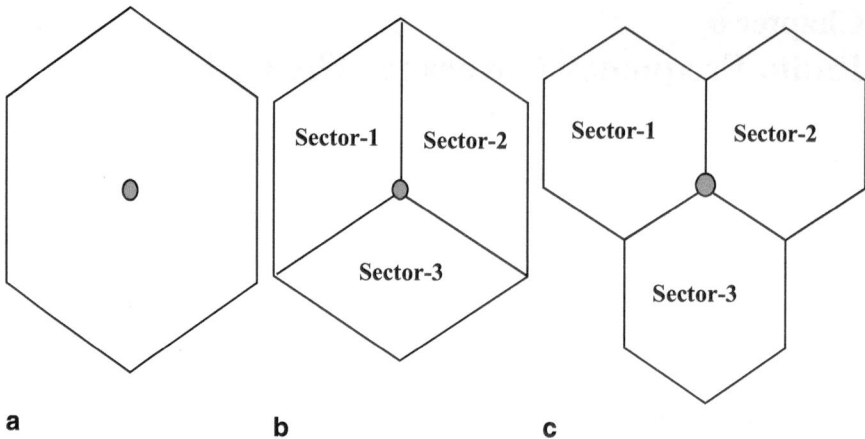

a   **b**   **c**

Fig. 6.1 Illustration of OMNI and sectorized cells. **a** OMNI cell, **b** three sectored cell, **c** sectorized cell (Tri-cellular plan)

antenna height, antenna gain, Propagation environment, pathloss characteristics and the signal strength at the cell edge. The signal strength, also known as Received Signal Level (RSL is the same all around the cell edge. These cells are distributed over land areas, offering services to tens of thousands of mobile phone users in a given service area. Also, each cellular base station is connected via the Network Operation Center, which provides today's land-mobile communication all over the world.

Since the RF environment is fuzzy and unpredictable, a practical cell is highly irregular and not useful for analytical purposes. An ideal cell, on the other hand, is a perfect sphere, which applies to free-space only. Considering all these shortcomings, the cellular industries have adopted the hexagonal cell as the basis for the analysis and design of all cellular systems. To illustrate, we Consider Fig. 6.2, where,

R   Cell Radius of the ideal cell (circle)
r   Effective radius of the analytical cell (center to vertex)

Using plane geometry, we obtain:

$$Cos(30^0) = \frac{r}{R} = \frac{\sqrt{3}}{2}$$ (6.1)

Solving for r,

$$r = \frac{\sqrt{3}}{2}R = 0.866R$$ (6.2)

We shall use this geometrical configuration for analysis throughout this chapter.

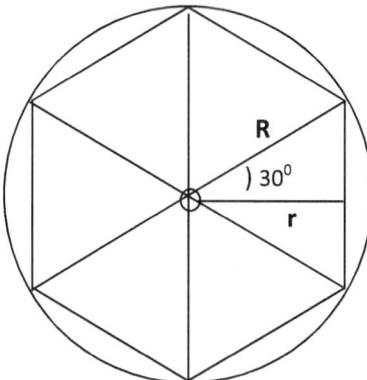

Fig. 6.2 The analytical hexagonal OMNI cell

6.3   The Distance Between Two Hexagonal Cells

The distance between two hexagonal cells (D) can be obtained by placing two hexagonal cells as shown in Fig. 6.3. This is given by,

$$D = 2r = 2\frac{\sqrt{3}}{2}R = 1.732R \tag{6.3}$$

The performance metric, known as D/R ratio, can be obtained from Eq. 6.3 as:

$$\frac{D}{R} = 1.732 \tag{6.4}$$

This parameter will be used to calculate carrier to interference ratio C/I, as we shall see later.

In Fig. 6.3, we also notice that two adjacent hexagonal cells are equivalent to two overlapping circles. This overlap region is the well-documented "Hand-Off" region.

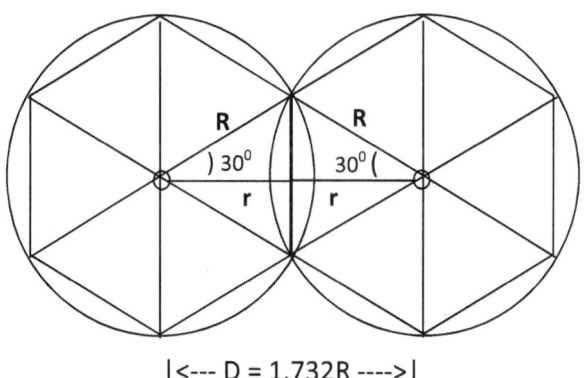

Fig. 6.3 Distance between two hexagonal cells

6.4   Frequency Reuse and C/I

6.4.1   C/I Due to a Single Interferer

In cellular communications, frequencies are reused in different cells, which mean that another mobile can use the same frequency, thereby causing co-channel interference or carrier to interference (C/I) [4–6]. As an illustration, we consider Fig. 6.3, where the same frequency is used in Cell-A and Cell-B. Therefore, a mobile communicating with Cell-A will also receive the same frequency from the distant Cell-B. This is analogous to the "Near-Far" problem, causing co-channel interference. We use the following method to determine this interference Fig. 6.4.

Let,

RSL_A Received signal level at the mobile from Cell-A
d_A Distance between the mobile and Cell-A
RSL_B The received signal level at the mobile from Cell-B
d_B Distance between the mobile and Cell-B
γ Pathloss exponent

Then we can write,

$$RSL_A \propto (d_A)^{-\gamma}$$
$$RSL_B \propto (d_B)^{-\gamma}$$

(6.5)

The ratio of the signal strengths at the mobile will be:

$$\frac{RSL_A}{RSL_B} = \left(\frac{d_A}{d_B}\right)^{-\gamma} = \left(\frac{d_B}{d_A}\right)^{\gamma}$$

(6.6)

In Eq. 6.6, RSL_A is the RF signal received from the serving cell. Therefore, this is the desired signal and we redefine this signal as the carrier signal C. We also assume

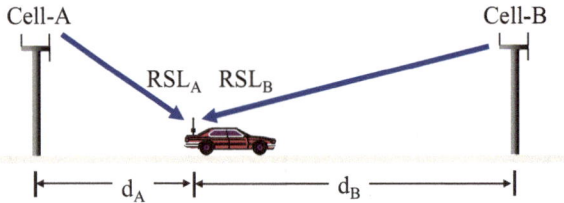

Fig. 6.4 Carrier to Interference ratio (C/I) due to a single interferer

that the mobile is at the cell edge from the serving cell-A, which is the cell radius R. On the other hand, RSL_B is the undesired signal received from $Cell_B$ and we redefine this signal as the interference signal I. The corresponding interference distance $d_B = D$; D being the reuse distance. Therefore, Eq. 6.6 can be written as a carrier to interference ratio (C/I), due to a single interferer, as:

$$\frac{C}{I} = \left(\frac{D}{R}\right)^\gamma \tag{6.7}$$

In decibel, it can be written as:

$$\frac{C}{I}(dB) = 10\text{Log}\left(\frac{D}{R}\right)^\gamma \tag{6.8}$$

6.4.2   C/I Due to Multiple Interferers

In hexagonal cellular geometry, each hexagonal cell is surrounded by six hexagons. Therefore, in a mature cellular system, there can be six primary interferers. The total interference from all six interferers will be

$$6RSL_B \propto (d_B)^{-\gamma}$$
$$RSL_B \propto \frac{1}{6}(d_B)^{-\gamma} \tag{6.9}$$

Therefore, the effective interference ratio is:

$$\frac{C}{I} = \frac{RSL_A}{RSL_B} = \frac{1}{6}\left(\frac{d_A}{d_B}\right)^{-\gamma} = \frac{1}{6}\left(\frac{d_B}{d_A}\right)^\gamma = \frac{1}{6}\left(\frac{D}{R}\right)^\gamma \tag{6.10}$$

And in decibel,

$$\frac{C}{I}(dB) = 10\text{Log}\left[\frac{1}{6}\left(\frac{D}{R}\right)^\gamma\right] \tag{6.11}$$

Therefore, by knowing the reuse distance, the C/I ratio can be determined. Or, by knowing the C/I requirement, the reuse distance can be determined in a given propagation environment. The reuse distance D can be determined from plane geometry and the cell radius can be obtained from the propagation model Fig. 6.5.

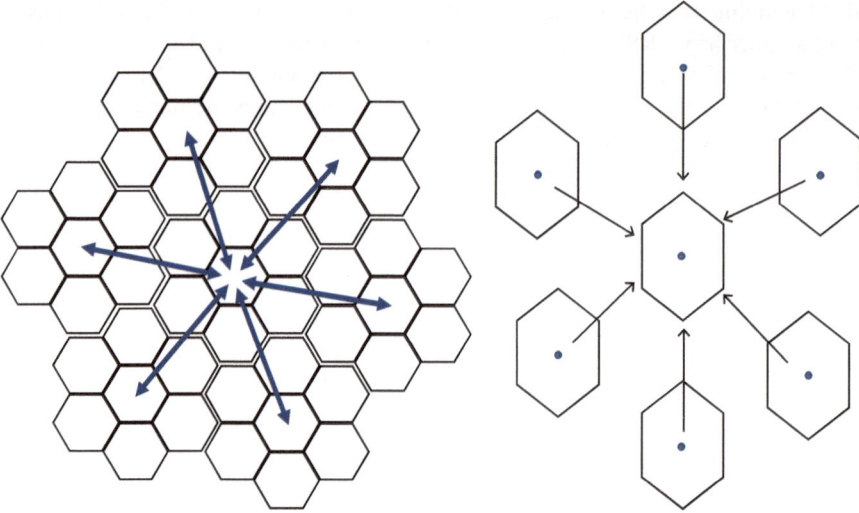

Fig. 6.5 C/I due to multiple interferers. Group of frequencies used in the center cell are reused in the surrounding six cells

6.5 Frequency Reuse

6.5.1 Basic Concept

Cellular communication is a multiple access system where several non-interfering channels are combined to form a channel group and assigned to a cell site. Since there are a limited number of channels, these channel groups are reused at a regular interval of distances. This is an important engineering task, which determines system capacity and performance.

Several frequency reuse techniques, generally known as frequency planning or channel assignment techniques, are available. Some of the most widely used frequency planning techniques are given below [5]:

- N=7 Frequency Reuse Plan
- N=3 Frequency Reuse Plan

The N-=7 is the classical cellular architecture, which is based on hexagonal geometry. It was originally developed by V.H. MacDonald in 1979. It ensures adequate channel reuse distance to an extent where co-channel interference is low and acceptable, while maintaining a high channel capacity. These frequency plans are briefly presented to illustrate the concept.

The scheme is shown in Fig. 6.6 where we have a cluster of 7 OMNI cells and a cluster of 7-sectorized cells. OMNI cell use OMNI directional (all directions) antennas. In the sectorized scheme, each cell is divided into three sectors, 120° each. Directional antennas are used in each sector. According to the art of channel assign-

OMNI   Sectored

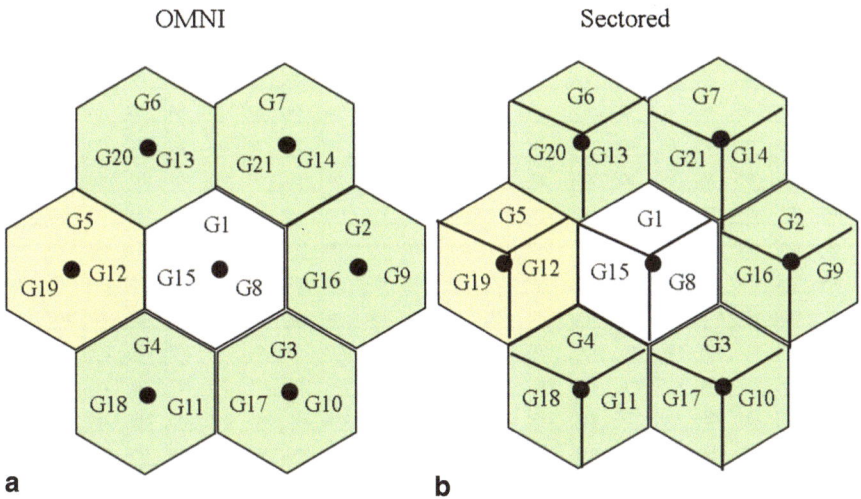

a   **b**

Fig. 6.6 The classical 7-cell cluster. **a** OMNI pattern and **b** 3-Sectored pattern.

ment technique, all the available channels are grouped into 21 frequency groups as follows:

G1, G2, G3, G4, G5, G6, G7, G8, G9, G10, G11, G12, G13, G14, G15, G16, G17, G18, G19, G20, G21.

Each frequency group has several frequencies (known as channels). These 21 frequency groups are equally distributed among the cells/sectors. Notice that each OMNI cell gets three frequency groups or a total of 21 frequency groups in the 7 cell cluster. On the other hand, a sectorized cell gets one frequency group per sector for a total of three frequency groups per cell. The total number of frequency groups per cluster is still the same as in the OMNI scheme.

6.5.2 Example of N = 7 OMNI Frequency Plan

In the $N = 7$ frequency reuse plan, the available channels are equally divided among 7 cells known as a 7-cell cluster. As an illustration, we will consider the classical North American 1G and 2G standards [1, 2]. According to these standards, the channel assignment is accomplished by forming 21 frequency groups per band (Band-A and Band-B) as shown in Tables 6.1 and 6.2 respectively [5]. Frequency assignment is then based on a distribution plan described earlier in the previous section. The corresponding $N = 7$ cell cluster is shown in Fig. 6.7 having three frequency groups per cell.

The total number of frequency groups per cluster is $7 \times 3 = 21$. Channel assignment is based on the following sequence: $(N, N + 7, N + 14)$ where N is the cell number $(N = 1, 2, 7)$. The channel-grouping scheme is shown in Table 6.3. Figure 6.7

Table 6.1 N = 7/21 A-Band Frequency Chart

1	2	3	4	5	6	7	8	9	10	11	12	13	14	15	16	17	18	19	20	21
22	23	24	25	26	27	28	29	30	31	32	33	34	35	36	37	38	39	40	41	42
43	44	45	46	47	48	49	50	51	52	53	54	55	56	57	58	59	60	61	62	63
64	65	66	67	68	69	70	71	72	73	74	75	76	77	78	79	80	81	82	83	84
85	86	87	88	89	90	91	92	93	94	95	96	97	98	99	100	101	102	103	104	105
106	107	108	109	110	111	112	113	114	115	116	117	118	119	120	121	122	123	124	125	126
127	128	129	130	131	132	133	134	135	136	137	138	139	140	141	142	143	144	145	146	147
148	149	150	151	152	153	154	155	156	157	158	159	160	161	162	163	164	165	166	167	168
169	170	171	172	173	174	175	176	177	178	179	180	181	182	183	184	185	186	187	188	189
190	191	192	193	194	195	196	197	198	199	200	201	202	203	204	205	206	207	208	209	210
211	212	213	214	215	216	217	218	219	220	221	222	223	224	225	226	227	228	229	230	231
232	233	234	235	236	237	238	239	240	241	242	243	244	245	246	247	248	249	250	251	252
253	254	255	256	257	258	259	260	261	262	263	264	265	266	267	268	269	270	271	272	273
274	275	276	277	278	279	280	281	282	283	284	285	286	287	288	289	290	291	292	293	294
295	296	297	298	299	300	301	302	303	304	305	306	307	308	309	310	311	312			
313	**314**	**315**	**316**	**317**	**318**	**319**	**320**	**321**	**322**	**323**	**324**	**325**	**326**	**327**	**328**	**329**	**330**	**331**	**332**	**333**
															667	668	669	670	671	672
673	674	675	676	677	678	679	680	681	682	683	684	685	686	687	688	689	690	691	692	693
694	695	696	697	698	699	700	701	702	703	704	705	706	707	708	709	710	711	712	713	714
715	716																			
									991	992	993	994	995	996	997	998	999	1000	1001	1002
1003	1004	1005	1006	1007	1008	1009	1010	1011	1012	1013	1014	1015	1016	1017	1018	1019	1020	1021	1022	1023

gives two versions of channel assignments. Notice that adjacent channels appear within the cluster, causing adjacent channel interference. It indicates that the total elimination of channel adjacency is practically impossible in the N = 7 plan, which gives rise to adjacent channel interference throughout the network.

6.5.3 Evaluation of Co-Channel Interference

Co-Channel Interference (CCI), also known as Carrier to Interference ratio (C/I, arises from multiple uses of the same frequency. For OMNI sites (see Fig. 6.6), C/I is given by [5, 6]:

$$\frac{C}{I} = 10 \log \left[\frac{1}{k} \left(\frac{D}{R} \right)^{\gamma} \right]$$

(6.12)

Where

k =   Number of Co-Channel Interferers

Table 6.2 N = 7/21 B-Band Frequency Chart

334	335	336	337	338	339	340	341	342	343	344	345	346	347	348	349	350	351	352	353	354
355	356	357	358	359	360	361	362	363	364	365	366	367	368	369	370	371	372	373	374	375
376	377	378	379	380	381	382	383	384	385	386	387	388	389	390	391	392	393	394	395	396
397	398	399	400	401	402	403	404	405	406	407	408	409	410	411	412	413	414	415	416	417
418	419	420	421	422	423	424	425	426	427	428	429	430	431	432	433	434	435	436	437	438
439	440	441	442	443	444	445	446	447	448	449	450	451	452	453	454	455	456	457	458	459
460	461	462	463	464	465	466	467	468	469	470	471	472	473	474	475	476	477	478	479	480
481	482	483	484	485	486	487	488	489	490	491	492	493	494	495	496	497	498	499	500	501
502	503	504	505	506	507	508	509	510	511	512	513	514	515	516	517	518	519	520	521	522
523	524	525	526	527	528	529	530	531	532	533	534	535	536	537	538	539	540	541	542	543
544	545	546	547	548	549	550	551	552	553	554	555	556	557	558	559	560	561	562	563	564
565	566	567	568	569	570	571	572	573	574	575	576	577	578	579	580	581	582	583	584	585
586	587	588	589	590	591	592	593	594	595	596	597	598	599	600	601	602	603	604	605	606
607	608	609	610	611	612	613	614	615	616	617	618	619	620	621	622	623	624	625	626	627
628	629	630	631	632	633	634	635	636	637	638	639	640	641	642	643	644	645	646	647	648
649	650	651	652	653	654	655	656	657	658	659	660	661	662	663	664	665	666			
				717	718	719	720	721	722	723	724	725	726	727	728	729	730	731	732	
733	734	735	736	737	738	739	740	741	742	743	744	745	746	747	748	749	750	751	752	753
754	755	756	757	758	759	760	761	762	763	764	765	766	767	768	769	770	771	772	773	774
775	776	777	778	779	780	781	782	783	784	785	786	787	788	789	790	791	792	793	794	795
796	797	798	799																	

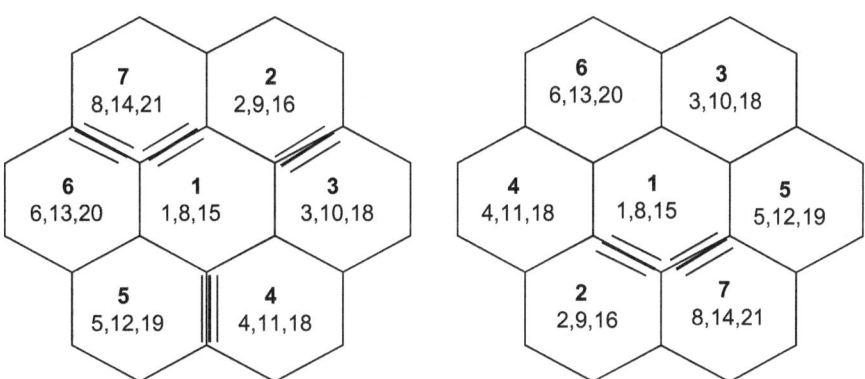

▬▬▬ Adjacent channel interference

Fig. 6.7 N = 7, Frequency reuse for OMNI plan showing re-appearance of adjacent channels

Table 6.3 N=7/21 OMNI, channel grouping

Cell Number	1	2	3	7	5	6	7
Freq. Group							
N	1	2	3	4	5	6	7
N+7	8	9	10	11	12	13	14
N+14	15	16	17	18	19	20	21

γ Propagation constant
D   Frequency reuse distance
R   Cell radius

The distance ratio (reuse distance) D/R is given by

$$\frac{D}{R} = \sqrt{3N} \tag{6.13}$$

For N=7 frequency plan

$$\frac{D}{R} = \sqrt{3x7} = 4.58 \tag{6.14}$$

The distance ratio can also be determined geometrically.

6.5.4   Evaluation of Adjacent Channel Interference

Adjacent channel interference arises from energy spillover between two adjacent channels. This can be evaluated with the aid of Fig. 6.8, where we assumed that the adjacent channel is assigned to the adjacent site and the ratio di/dc varies as the mobile moves towards or away from the cell. Moreover, the out-of-band signals are also assumed attenuated by the post-modulation filter at least by 26 dB (EIA Standard) [1, 2]. Then the Adjacent Channel Interference will be:

$$ACI = -10 \log \left[\left(\frac{d_i}{d_c} \right)^{\gamma} \right] + Adj.Ch.Isolation \tag{6.15}$$

Where the adjacent channel isolation is provided by the post-modulation filter, which is generally ≈ 26 dB. As an example, an adjacent channel in the in the adjacent site has di/dc=1, resulting in ACI ≈ -26 dB. An adjacent channel in an alternate channel (one cell away) provides much greater than 26 dB isolation due to a higher distance ratio (di/dc > 1).

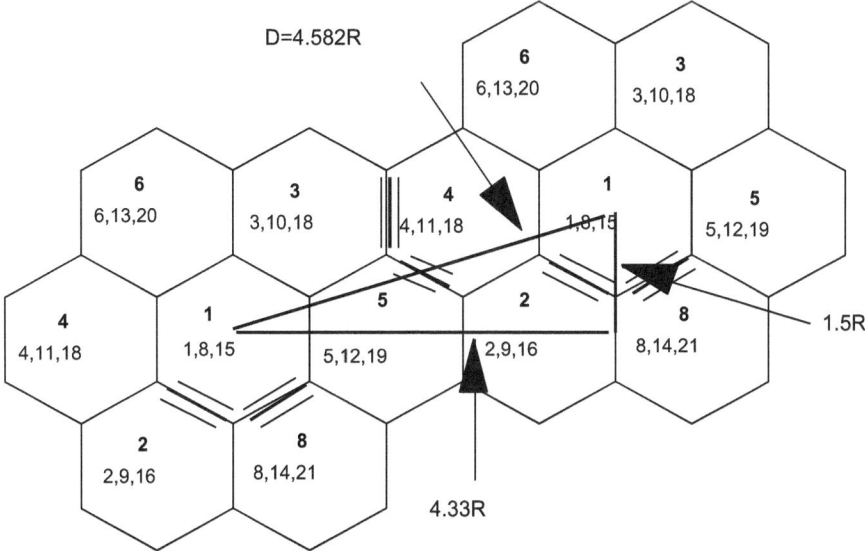

Fig. 6.8 Adjacent channel interference evaluation scheme

6.6   120° Sectorization

6.6.1   Basic Concept

The 120° Sectorization is achieved by dividing a cell into three sectors, 120° each, as shown in Fig. 6.9a. Each sector is treated as a logical OMNI cell, where directional antennas are used in each sector for a total of three antennas per cell. Figure 6.9b shows an alternate representation, which is known as Tri-Cellular plan [5, 6]. Both configurations are conceptually identical while the latter is convenient for channel assignment. Each sector uses one control channel and a set of different voice channels. Adequate channel isolations are maintained within and between sectors in order to minimize interference. This is attributed to channel assignment techniques, as we shall see later in this chapter.

Because directional antennas are used in sectored cells, it allows reuse of channels more frequently, thus enhancing channel capacity. Moreover, multipath components are also reduced due to directionalization, hence enhancing the C/I performance.

6.6.2   N = 7/21, 120° Sectorization Plan

The N = 7/21, 120° sectorization plan is based on the distribution of one frequency group per sector, three frequency groups per cell for a total of 21 frequency groups per cluster. This is shown in Fig. 6.10a for the conventional plan and in Fig. 6.10b

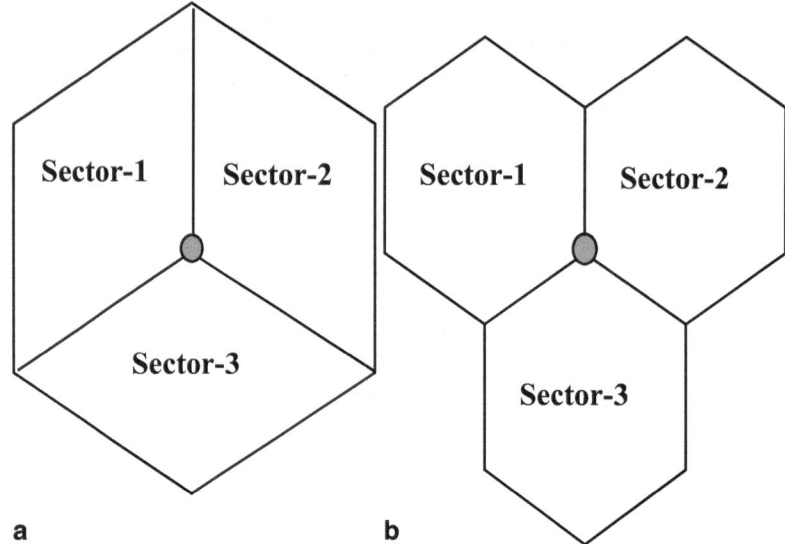

a   b

Fig. 6.9 Illustration of 120° sectorization. Directional antennas are used in each sector **a** Conventional representation **b** an alternate representation

for the tri-cellular plan where the sector is represented by hexagon. Channel distribution is based on N, N+7, N+14 scheme where N = 1, 2,.., 7. Therefore for N = 1, cell-1 uses frequency group-1 for sector-1, frequency group-8 for sector-2 and frequency group-15 for sector-3. Similarly, cell-2 uses frequency group 2, 9 and 16 for sector 1, 2 and 3 respectively.

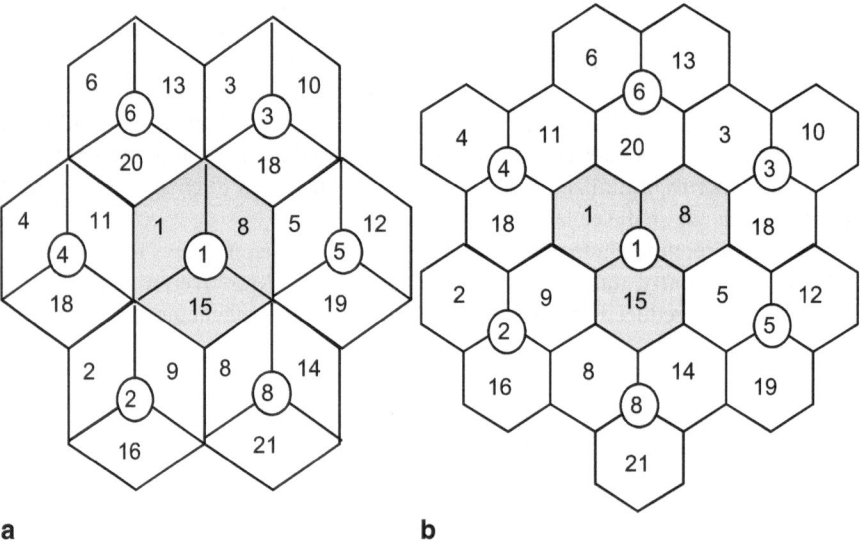

a   b

Fig. 6.10 a N = 7/21 sectorized configurations based on 120°. Sectorized plan. **b** N = 7/21 tri-cellular plan

6.6.3   N=7/21, 120-Deg. Co-Channel Interference

The 120-deg sectorization is achieved by dividing a cell into 3 sectors, while directional antennas are used in each sector. Thus antenna configuration and their directivity play an important role in determining the C/I performances. In order to illustrate this further, let us consider the diagram as shown in Fig. 6.11, where directional antennas are used for the present analysis. Antenna down tilt is also provided for an additional isolation, which must be taken into account.

The angle of down tilt is related to the cell radius and antenna height. This can be calculated by using plane geometry. It is given by the following equation:

$$R = \frac{H}{\tan \theta} \qquad (6.16)$$

Where,

R = Cell radius
H = Antenna height
θ = Angle of antenna down tilt

These assumptions modify the C/I prediction equation as

$$\frac{C}{I} = 10 \log \left[\frac{1}{k} \left(\frac{D}{R} \right)^{\gamma} \right] + \Delta dB \text{(due to antenna down tilt)} \qquad (6.17)$$

With k=6, γ=3.84, D/R=4.58 and ΔdB≈6 dB, we now obtain

$$\frac{c}{I} \approx 23.6 dB \qquad (6.18)$$

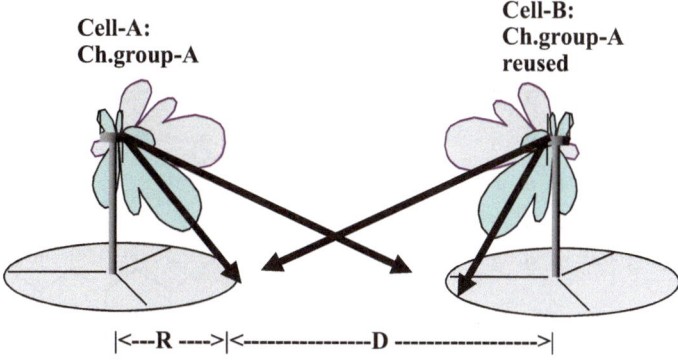

Cell-A:
Ch.group-A

Cell-B:
Ch.group-A
reused

|<---R ---->|<----------------D ------------------>|

Fig. 6.11 Illustration of antenna down tilt

The performance can be further improved by using antenna having a narrow vertical beam width.

6.7   N = 3 Tri-Cellular Plan

6.7.1   Alternate Channel Assignment

The N = 3 tri-cellular plan is based on a cluster of three tri-cells for a total of 9 logical cells [5]. Channel assignment is based on a 3×3 array of 9 frequency groups, distributed alternately among 9 logical cells as shown in Fig. 6.12. Because of alternate channel assignment, this arrangement completely eliminates adjacent channels from adjacent sites, thus reducing adjacent channel interference.

The growth plan is based on repetition of vertical and horizontal patterns in sequence as shown in Fig. 6.13. As can be seen, adjacent channel isolation is maintained throughout the network. It may be noted that only vertical and horizontal

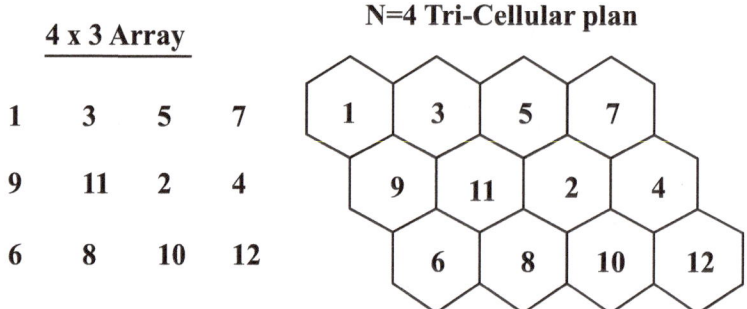

Fig. 6.12 The N = 3, tri-cellular plan having 9 logical cells as sectors

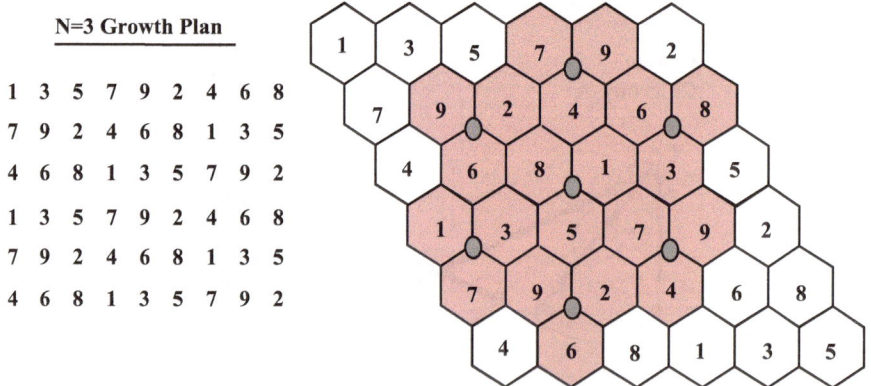

Fig. 6.13 N = 3, Tri-Cellular growth plan showing N = 7 mapping

expansions are possible in this scheme; this is due to the rhombic pattern of the cluster. The N=7 mapping is also evident in Fig. 6.13, which can be shown to expand throughout the network.

6.7.2   N=3 Cyclic Distribution of Channels

In this plan, the channel distribution is based on $(N, N+3, N+6)$ where $N=1,2,3$, N being the cell number. Using this scheme, we obtain

$$
\begin{array}{c}
\text{Cell-1 frequency group: 1, 4, 7} \\
\text{Cell-2 frequency group: 2, 5, 8} \\
\text{Cell-3 frequency group: 3, 6, 9}
\end{array} \tag{6.19}
$$

The corresponding N=3 cell cluster is given in Fig. 6.14, where the growth plan is already established. For example, if cell-1 is assumed to be in the center, the surrounding pattern would be 2, 3, 2, 3, 2, 3. Similarly, if cell-2 is in the center, the surrounding pattern would be 1, 3, 1, 3, 1, 3. Likewise, if cell-3 is in the center, the surrounding pattern would be 1, 2, 1, 2, 1, 2. This is shown in Fig. 6.15 where each distribution pattern is free of adjacent channels.

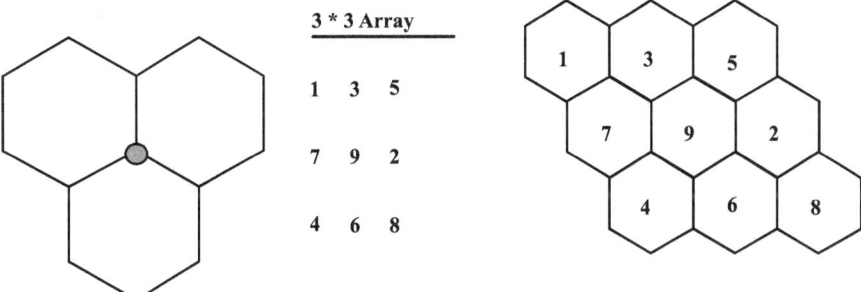

Fig. 6.14 N=3 tri-cellular plan

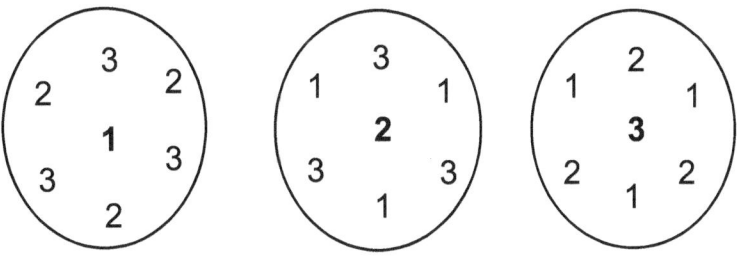

Fig. 6.15 Principle of cyclic distribution

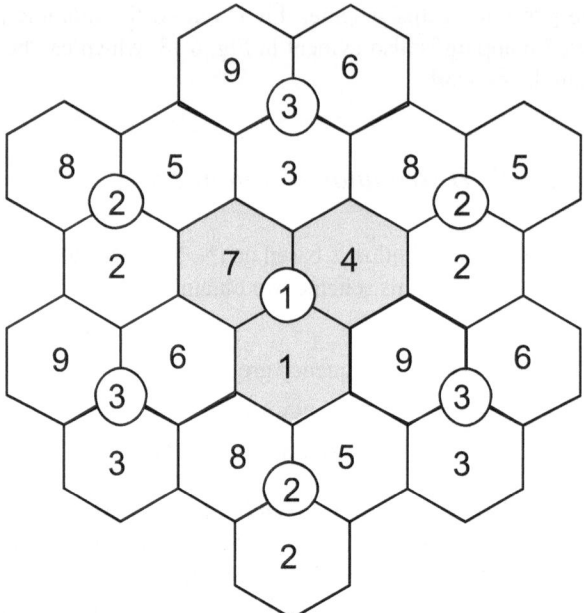

Fig. 6.16 N=3 growth plan showing N=7 mapping

Figure 6.16 shows a complete distribution pattern for case-1 where cell-1 is assumed to be in the center. we also see a mapping similarity between N=3 plan and N=7 plan. Therefore existing N=7 cell sites can be easily translated into N=3 plan simply by channel reassignment. Other distribution patterns can be obtained from the distribution principle according to Fig. 6.15.

6.7.3   N=3 Co-Channel Interference

Because of antenna directivity, it can be shown that there are three effective interferers in this scheme.

Therefore the Co-Channel Interference can be estimated as:

$$\frac{C}{I} \approx 10 \log \left[\frac{1}{3} \left(\frac{D}{R} \right)^{\gamma} \right] + \Delta dB(ave.) \tag{6.20}$$

$$\frac{D}{R} = \sqrt{3 \, x \, 3} = 3$$

With $\gamma \approx 3.84$, D/R=3 and $\Delta dB(ave.) \approx 6$ dB we get

$$\frac{C}{I} \approx 13.55 + 6 = 19.55 dB \tag{6.21}$$

Problem 6.1

Given: The received signal strength at the cell edge: RSL (dBm)=−90 dBm

Find: The corresponding RSL in watts

Solution 6.1

RSL (dBm)=10 Log [RSL(mW)=−90 dBm

Log [RSL (mW)=−90/10=−9 mW

[RSL (watts)=10^{-12} Watts

[NOTE: 1 milli watt=10^{-3} Watts]

Problem 6.2

Given: The received signal strength at the cell edge: RSL (W)=10^{-12} W

Find: The corresponding RSL in dBm.

Solution 6.2

$$RSL(dBW) = 10 \text{ Log} [RSL(W)$$
$$= 10 \text{ Log} (10-12)$$
$$= -120 \text{ dBW}$$
$$= -90(dBm)$$

[NOTE: 0 dBW=30 dBM]

Problem 6.3

Show that 0 dBW=30 dBm.

Solution 6.3

1 W=1000 mW

Therefore, in decibel:

10 Log (1 W)=10 Log (1000 mW)

Or

0 dBW=30 dBm.

Problem 6.4

Given:

- N=7 OMNI and sectorized frequency reuse plans
- Pathloss slope γ=4

Find:

a. The carrier to interference ratio C/I for the OMNI Plan.
b. The carrier to interference ratio C/I for the sectorized Plan.

Solution 6.4

a. For the OMNI plan:

$$\frac{C}{I} = 10\log\left[\frac{1}{k}\left(\sqrt{3xN}\right)^{\gamma}\right]$$

With N=7 OMNI, k=6. Using γ=4, the C/I becomes,

$$\frac{C}{I} = 18.6 \text{ dB} \tag{6.22}$$

b. For the N=7 sectorized plan, the C/I is written as:

$$\frac{C}{I} \approx 10\log\left[\frac{1}{k}\left(\sqrt{3xN}\right)^{\gamma}\right] + \Delta dB(ave.)$$

Where k=3. ΔdB is an additional dB margin due to directional antennas with down tilt. This value ranges from 6 to 8 dB depending on bean width and amount of down tilt. Using γ=4 for dense urban environment and ΔdB=6 dB, the C/I can be derived from the previous problem. This is given by:

C/I (N=7, Sectorized)=18.6 dB+6 dB=24.6 dB.

Problem 6.5

Given:

* N=3 OMNI and sectorized frequency reuse plan
* Pathloss slope γ=4

Find:

a. The carrier to interference ratio C/I for the OMNI Plan.
b. The carrier to interference ratio C/I for the sectorized Plan.

Solution 6.5

a. For the OMNI plan:

$$\frac{C}{I} = 10\log\left[\frac{1}{k}\left(\sqrt{3xN}\right)^{\gamma}\right]$$

With N=3 OMNI, k=6. Using γ=4, the C/I becomes,

$$\frac{C}{I} = 14 \text{ dB} \tag{6.23}$$

b. For the N=3 sectorized plan, the C/I is written as:

$$\frac{C}{I} \approx 10 \log\left[\frac{1}{k}\left(\sqrt{3xN}\right)^\gamma\right] + \Delta dB(ave.)$$

Where k=3. ΔdB is an additional dB margin due to directional antennas with down tilt. This value ranges from 6 to 8 dB depending on bean width and amount of down tilt. Using N=3, γ=4 for dense urban environment and ΔdB=6 dB, the C/I can be derived from the previous problem. This is given by:

C/I (N=3, Sectorized)=14 dB+6 dB=20 dB.

6.8 Conclusions

- Discussed radio Frequency coverage and provided the concept of cell
- Rationalized the use of hexagonal cell geometry and calculated cell radius
- Provided the concept of OMNI and Sectorized cells
- Provided the concept of Cell cluster
- Presented N=7 frequency reuse plan and carrier to interference ratio (C/I)
- Presented N=3 frequency reuse plan and carrier to interference ratio (C/I)
- Discussed the benefit of antenna down tilt and calculated the down tilt angle

References

1. IS-54, "Dual-Mode Mobile Station-Base Station Compatibility Standard EIA/TIA Project Number 2215", December, 1989.
2. IS-95 "Mobile Station—Base Station Compatibility Standard for dual Mode Wide band Spread Spectrum Cellular Systems," TR 45, PN-3115, March 15. 1993.
3. V.H. Mac Donald, "The Cellular Concept", "The Bell System Technical Journal", Vol. 58, No. 1, January 1979.
4. W.C.Y. Lee, "Mobile Cellular Telecommunications Systems", McGraw-Hill Book Co. N.Y., 1989
5. Saleh, Faruque ";Cellular, Mobile Systems Engineering";, Artech, House, Boston, 1996.
6. Saleh Faruque, "Directional Pseudonoise Offset Assignment in a CDMA Cellular Radio Telephone System, US Patent No. 5883889", Granted March 16, 1999.

Chapter 7
Global RF & Co$_2$ Pollution: Transition to Green Cellular Technology

7.1 Introduction to Global RF and CO$_2$ Pollution

Cell phone towers emit radio frequencies (RF) 24 hours a day and we absorb them continuously. We also absorb RF from the cell phone when we use it, indicating possible public health issues [1]. These cell phone towers, owned and operated by service providers, use power 24 hours a day. Consequently, power for this technology also contributes to global CO$_2$ pollution, which is expected to rise because of current explosion in wireless data [2, 3]. Almost all countries around the globe face this dilemma, giving rise to ever-increasing Global RF & CO$_2$ pollution.

Transition to green cellular networks could be an effective solution to these problems. This chapter aims to develop the concept of Green Wireless Technology to conserve energy, as well as to reduce RF & CO$_2$ pollution. First, we will examine Electron Spin Resonance (ESR) [4] and determine the range of frequencies we absorb. This will enable us to identify those frequencies we absorb the most and take appropriate measures to allocate cellular spectrum. Next, the amount of power that we absorb will be estimated by means of drive test data, where the received power will be measured as a function of distance. This will enable us to allocate link budget in a particular cell site, while keeping the radiation level to an acceptable limit. Finally, we will learn that in every propagation environment, there exists a free space propagation medium, due to the existence of Fresnel Zones. This gives rise to Fresnel zone break point d_o, where d_o is the distance between the transmitter and the receiver [5–7]. Radio frequency propagation within d_o is similar to free space propagation. This will enable us to design low power small cells to conserve energy, thereby reducing RF & CO2 pollution.

With this foundation in mind, we will develop a hierarchical cellular structure for the next generation green cellular network comprising Macrocell, Microcell, Picocell and Femtocell.

© Springer International Publishing Switzerland 2015   89
S. Faruque, *Radio Frequency Propagation Made Easy,* SpringerBriefs in Electrical
and Computer Engineering, DOI 10.1007/978-3-319-11394-4_7

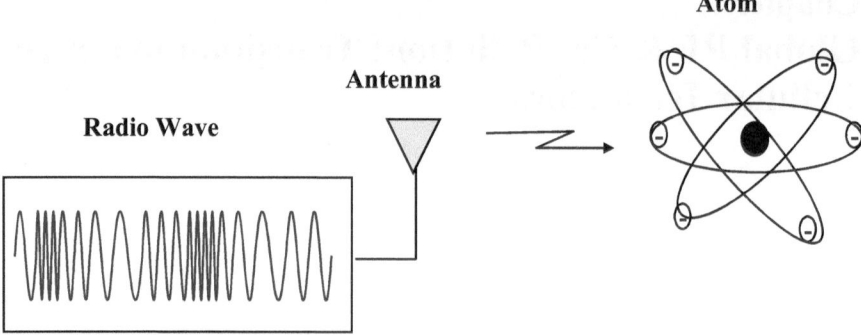

Fig. 7.1 Electron spin resonance

7.2   Mechanism of RF Absorption

7.2.1   The Range of Frequencies We Absorb

In this section we will examine the phenomenon known as Electron Spin Resonance (ESR) [4], also known as "Bosons", after the famed physicist Dr. S. Bose. Dr. Bose said, when an atomic particle is irradiated by means of an electromagnetic wave, the atom absorbs electromagnetic energy at a frequency determined by the atomic number. This gives rise to Electron Spin Resonance (ESR). This is conceptually shown in Fig. 7.1. This theory is the basis of the modern microwave oven (2.4 GHz). This is the frequency that causes molecular friction, generating heat. Since human bodies contain water, sugar, salt etc., an instrument has been developed to examine RF absorption in water with other impurities added in.

7.2.2   An Instrument to Detect RF Absorption

The instrument is based on HP8720A network analyzer as shown in Fig. 7.2. The unknown sample is placed between the transmit and the receive antenna, where the antenna separation is less than the Fresnel Zone break point (see chap. 3). As a result, multipath components will be cancelled and the instrument will exhibit microwave absorption only due to ESR, as long as the antenna separation is less than the Fresnel zone break point d_o, where,

$$d_o = \frac{4h_1h_2}{\lambda} = \frac{4h_1h_2f}{c} \qquad (7.1)$$

Fig. 7.2 The instrument for measuring RF absorption

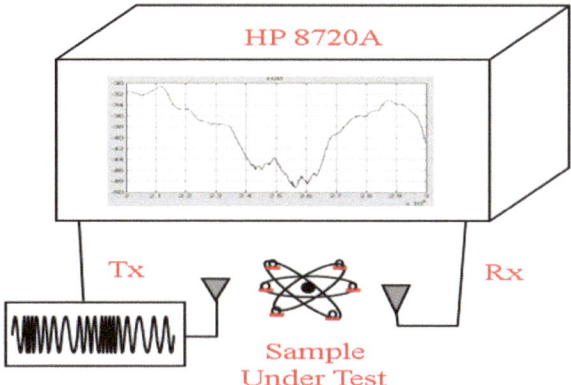

f   frequency
c   velocity of light
h_1 Transmit (Tx) antenna height
h_2 Receive (Rx) antenna height,

The distance (d_0) is known as the Fresnel zone break point which is proportional to frequency and antenna height as shown in Fig. 3.4. The line of sight pathloss slope within d_0 is similar to free space pathloss since diffraction and multipath phenomenon generally occurs beyond this region

7.2.3   Amount of Power We Absorb

The amount of power that we absorb can be estimated by means of drive test and data collection, where the received power is measured as a function of distance. This is a standard procedure in the cellular industry to optimize cell phone networks. The result is shown in Fig. 7.3 for a typical urban environment.

Here, we see that the received signal level at 1 km from the tower is approximately -60 dBm, which corresponds to one microwatt. We also notice that the received signal level decays logarithmically as a function of distance. At 5 km away from the cell phone tower, the received signal level is approximately in the range of -80 dBm. While this is a very tiny amount coming from a single channel, this goes on 24 hours a day and there are numerous multi-channel cell phone towers around us, and we absorb them all 24/7. Moreover, each base station as well as cell phones is responsible for RF and CO_2 pollution. Therefore, a possible solution to these problems would be to reduce power consumption, thereby reducing RF and CO_2 pollution.

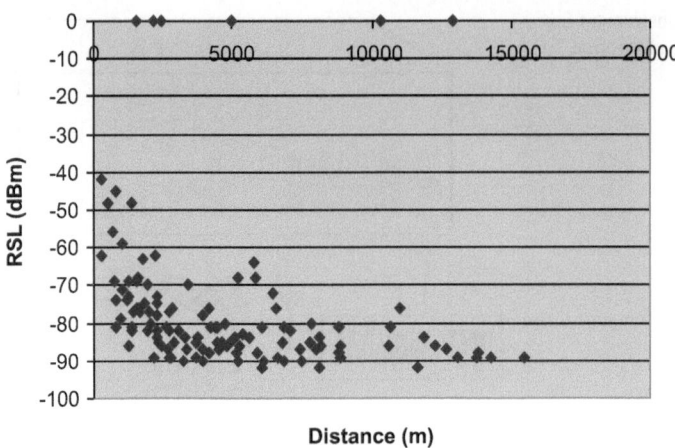

Fig. 7.3 Received power as a function of distance

7.2.4   Energy Efficient Cell Design

Energy efficient cells can be readily obtained from the drive test data given in Fig. 7.3. Here we have plotted RSL as a function of distance, where the distance is measured in meter and RSL is measured in dBm. We will use the following design problems to illustrate the design process.

Problem 7.1: Macro Cell Design from Drive Test Data (Fig. 7.3)
Given:

- RSL at the cell edge: RSL(dBm) = 90 dBm
- Standard deviation = 10 dB
- Confidence level = 90 % (Use Fig. 7.4)

Find:

a. The new RSL in dBm having 90 % confidence level (Use Fig. 7.4)
b. The new cell radius

Solution 7.1:

a. With 90 % confidence level, we have (from Fig. 7.4):
   Z = 1.3. The new RSL in dBm = σz + RSL (old)
   = 10 × 1.3–90 dBm
   = −77 dBm
b. The cell radius corresponding to − 77 dBm at the cell edge can be readily obtaind from Fig. 7.3, which is approximately 5 km. This is the macro cell as shown in Fig. 7.5.

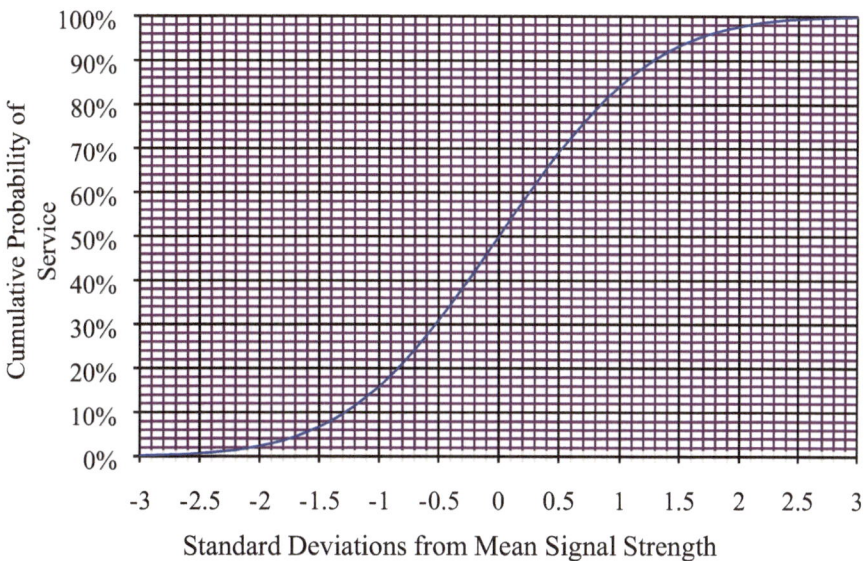

Fig. 7.4 Cumulative distribution

Fig. 7.5 Conversion of a
macro cell into a micro cell.
Macro cell radius = 5 km,
Micro cell radius = 1 km.
Power saving is approxi-
mately 200 times. Confidence
level = 90 % for both cells

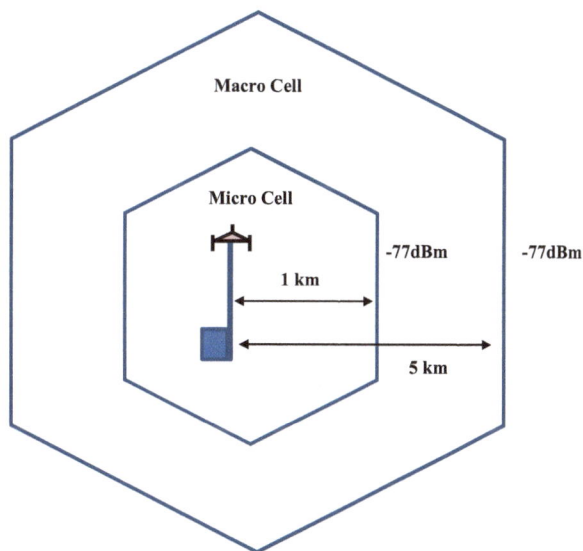

Problem 7.2: Micro Cell Design from Macrocell
Given:

- Macro cell radius = (5 km)
- RSL at the Macro-cell edge = −77 dBm
- The desired Micro-cell radius is 1 km,

Find:

a. The RSL at the Micro cell edge in dBm
b. Estimate the amount of power saved

Solution 7.2:

a. From Fig. 7.3, the value of RSL at 1 km is approximately −57 dBm, which is 77−57=20 dB strong.
b. Since the signal strength at the Micro cell edge is 200 times strong, the transmitter power can be reduced by 20 dB or 200 times to become −77 dBm at 1 km. This is the micro cell as shown in the inset of Fig. 7.5.

Problem 7.3
Briefly explain what you have learned from the above two problems.
Answer to problem 7.3:

- The transmit power reduces logarithmically as the cell radius reduces linearly.
- RF and CO$_2$ pollution reduce accordingly.
- Therefore, low power cell (Green Cell) design is essential to reduce Global RF and CO$_2$ pollution.

7.3　Green Cellular Technology

7.3.1　Background & Definition

Cell phone towers emit radio frequencies (RF) 24 hours a day and we absorb them continuously. We also absorb RF from the cell phone when we use it, indicating possible public health issues. Also, power for this technology is responsible for CO$_2$ emission, which is expected to rise due to the current explosion in wireless data. Almost all countries around the globe face this dilemma, giving rise to ever-increasing Global RF& CO$_2$ pollution. Transition to green cellular networks could be an effective solution to these problems. Therefore, green cellular technology can be defined as the one that reduces power consumption in cellular networks, thereby mitigating RF and CO$_2$ pollution.

This section aims to develop the concept of Green Wireless Technology to conserve energy, as well as to reduce RF & CO$_2$ pollution. We will learn that in every propagation environment, there exists a free space propagation medium, due to the

Fig. 7.6 Illustration of Fresnel zone break point

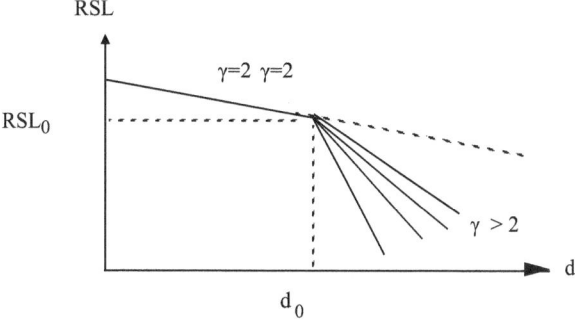

existence of Fresnel Zones. This gives rise to Fresnel zone break point d_o, where d_o is the distance between the transmitter and the receiver. Radio frequency propagation within d_o is similar to free space propagation. This will enable us to design low power small cells to conserve energy, thereby reducing RF & CO_2 pollution. With this foundation in mind, we will develop a hierarchical cellular structure for the next generation green cellular network comprising Macrocell, Microcell, Picocell and Femtocell.

7.3.2   Fresnel Zone Break Point & Cell Radii

In every propagation environment, there exists a free space propagation medium due to the existence of Fresnel Zones [5]. This gives rise to Fresnel zone break point d_o, where d_o is the distance between the transmitter and the receiver. Radio frequency propagation within do is similar to free space and the pathloss slope $\gamma = 2$, i.e., square law attenuation (see chap. 2–3).

For outdoor environment, this break point is given by the following equation [5]:

$$d_o = \frac{4h_1 h_2}{\lambda} = \frac{4h_1 h_2 f}{c} \tag{7.2}$$

Where,
h_1 Transmit antenna height
h_2 Receive antenna height
f Frequency
c velocity of light

This distance (d_o) is known as the Fresnel zone break point (do), which is proportional to frequency and antenna height as shown in Fig. 7.6. The line of sight pathloss slope within d_o is similar to free space pathloss since diffraction and multipath phenomenon generally occurs beyond this region.

Fig. 7.7 Three-ray indoor mode l [5]

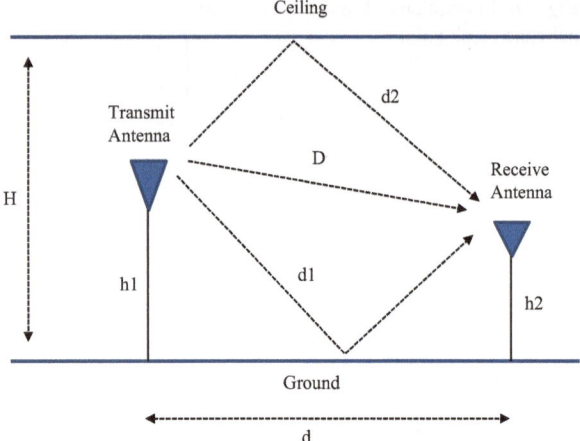

In-building coverage is given by a three-ray model, where the antenna is located within the building having ground reflections as well as reflections from the ceiling. This is shown in Fig. 7.7 where

H   Ceiling height
h$_1$ Transmit antenna height
h$_2$ Receive antenna height
d   Antenna separation
D   Direct path
d$_1$ Ground reflected path
d$_2$ Ceiling reflected path

The break point is given by the following equation [5]:

$$d_o = d = \frac{4(H - h_2)h_2}{\lambda} = \frac{4(H - h_2)h_2 f}{c} \tag{7.3}$$

The break point (d$_o$) is proportional to the frequency.

In tunnels and subways, there exists a Fresnel Zone break point, where the direct path and the reflected path are exactly 180° out of phase. RF signals within this distance are in-phase and do not cancel each other out. On the other hand, RF signals beyond this point are out of phase and suffer from multipath cancellations. Therefore, by knowing the frequency and the geometry of the device, the RF transmit and receive antennas can be properly positioned to create a free-space propagation region between the antennas as shown in Fig. 7.8. It is assumed that for every reflected path there is an identical path from the opposite wall inside the cylindrical tube. So that the path differences are identical, The break point is given by the following equation:

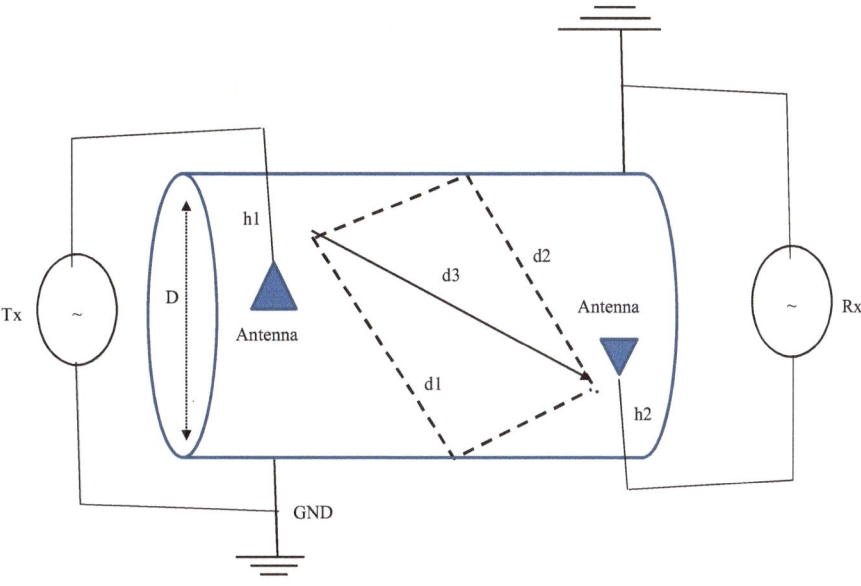

Fig. 7.8 RF propagation in tunnels and subways

$$d = d_o = \frac{4(D-h_1)(D-h_2)}{\lambda} = \frac{4(D-h_1)(D-h_2)f}{c} \qquad (7.4)$$

Where,

D   Diameter of the tube,
h_1 Transmit antenna height,
h_2 Receive antenna height and
d_o Antenna separation,
f   frequency.

Therefore, the break point is proportional to the frequency.

Therefore we can conclude that there exists a free-space pathloss within a circular tube, provided the antenna separation is less than the Fresnel zone break point d_o.

7.3.3   The Green Cell

Figure 7.9 shows the desired green cell (Inner cell). Here, the inner cell radius falls within the Fresnel zone break point, which gives rise to free space propagation within the cell resulting in $\gamma = 2$. On the other hand, the outer cell radius falls beyond the Fresnel zone break point, giving rise to terrestrial propagation. This implies that the pathloss slope within the outer cell is greater than 2 ($\gamma > 2$). The break point

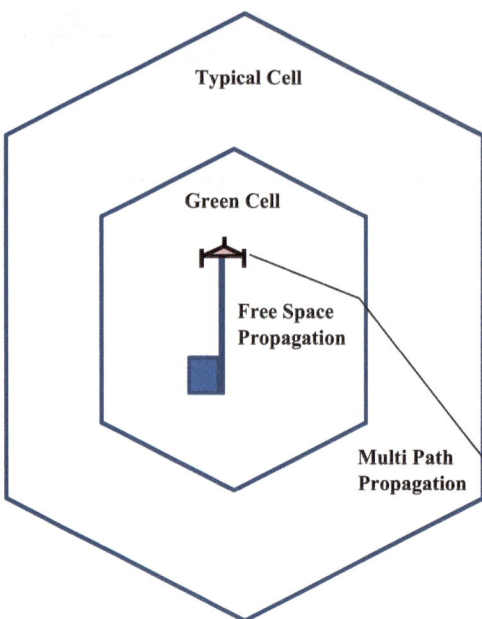

Fig. 7.9 Illustration of free space propagation medium in terrestrial environment. The radius of the inner cell is within the Fresnel zone break point and the outer cell is beyond the Fresnel zone break point

depends on antenna location, antenna height, velocity of light and frequency. Therefore, the cell radius will vary accordingly.

7.3.4   Green Cell Reuse & C/I

In cellular communications, C/I arise due to frequency reuse (see chap. 6). In this chapter we will examine this by means of Fig. 7.10, where Cell-A and Cell-B are Fresnel cells, i.e., the path loss slope within these cells follow square law attenuation. According to frequency reuse technique, Same frequency is used in Cell-A and Cell-B. Therefore, a mobile communicating with Cell-A will also receive the same frequency from the distant Cell-B, causing co-channel interference (C/I).

We use the following method to determine this interference.

Let,

RSL_A	Received signal level at the mobile from Cell-A
d_A	Distance between the mobile and Cell-A
γ_A	Pathloss exponent from cell B.
RSL_B	The received signal level at the mobile from Cell-B
d_B	Distance between the mobile and Cell-B
γ_B	Pathloss exponent from cell B.

Fig. 7.10 Carrier to Interference ratio (C/I) due to Fresnel zone break point

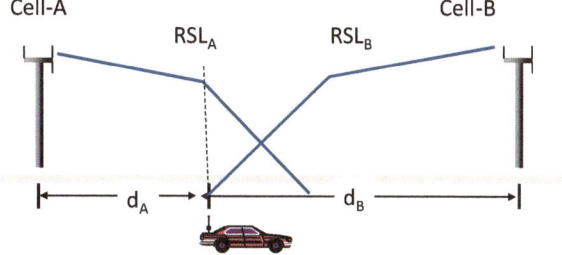

Then we can write,

$$RSL_A \propto \left(d_A\right)^{-\gamma A}$$
$$RSL_B \propto \left(d_B\right)^{-\gamma B} \tag{7.5}$$

The ratio of the signal strengths at the mobile will be:

$$\frac{RSL_A}{RSL_B} = \left(\frac{\left(d_A\right)^{-\gamma A}}{\left(d_B\right)^{-\gamma B}}\right) \tag{7.6}$$

In Eq. 7.6, RSL_A is the RF signal received from the serving cell. Therefore, this is the desired signal and we redefine this signal as the carrier signal C. We also assume that the mobile is at the cell edge from the serving cell-A, which is the cell radius R. On the other hand, RSL_B is the undesired signal received from $Cell_B$ and we redefine this signal as the interference signal I. The corresponding interference distance $d_B = D$; D being the reuse distance. Therefore, Eq. 7.6 can be written as a carrier to interference ratio (C/I), due to a single interferer, as:

$$\frac{C}{I} = \left(\frac{\left(d_A\right)^{-\gamma A}}{\left(d_B\right)^{-\gamma B}}\right) \tag{7.7}$$

In decibel, it can be written as:

$$\frac{C}{I}(dB) = 10 Log\left(\frac{\left(d_A\right)^{-\gamma A}}{\left(d_B\right)^{-\gamma B}}\right) \tag{7.8}$$

For multiple interferers, the above equation can be written as,

$$\frac{C}{I}(dB) = 10 Log\left(\frac{1}{k}\frac{\left(d_A\right)^{-\gamma A}}{\left(d_B\right)^{-\gamma B}}\right) \tag{7.9}$$

Where,

- k = Number of interferers (k = 6 for OMNI and k = 3 for sectorized cells)
- $\gamma_B > \gamma_A$. (γ_A. ~ 2, $\gamma_B > 2$)

Consequently, the C/I margin will be better than the typical reuse technique. We also notice that C/I increase rapidly as a function of reuse distance due to the difference in path loss slope. This makes Fresnel cells more attractive for energy efficient cellular networks.

In cellular communication, it is also a common practice to express C/I as a function D/R ratio as follows:

$$D = R\sqrt{3N} \tag{7.10}$$

Where

N   Frequency reuse factor (e.g., N = 7, 3 etc.)
D   Frequency reuse distance (d_B)
R   Cell radius (d_A)

With D = d_B and R = d_A, the C/I can be expressed as follows:

$$\frac{C}{I}(dB) = 10Log\left(\frac{1}{k}\frac{(R)^{-\gamma_A}}{\left(R\sqrt{3N}\right)^{-\gamma_B}}\right) \tag{7.11}$$

7.4   Green Cellular Hierarchy

7.4.1   Background

It was shown that, in every propagation environment, there exists a free space propagation medium due to the existence of Fresnel Zones. This gives rise to Fresnel zone break point d_o, where d_o is the cell radius. Radio frequency propagation within do is similar to free space and the pathloss slope $\gamma = 2$, i.e., square law attenuation (see chap. 2–3). Since the Fresnel zone break point depends on the operating frequency and antenna height, the radius of the cell will vary according to base station location and the transmit frequency. We will use these parameters to define Micro, Pico and Femto cells.

Figure 7.11 shows a conceptual green cellular hierarchy. It begins with the classical Macrocell followed by the derivation of Micro, Pico and Femto cells. The micro cell radius falls within the Fresnel zone break point, which gives rise to free space propagation within the cell resulting in $\gamma = 2$. On the other hand, the outer cell radius falls beyond the Fresnel zone break point, giving rise to terrestrial propagation. This implies that the pathloss slope within the outer cell is greater than 2 ($\gamma > 2$).

Fig. 7.11 Green cell
hierarchy

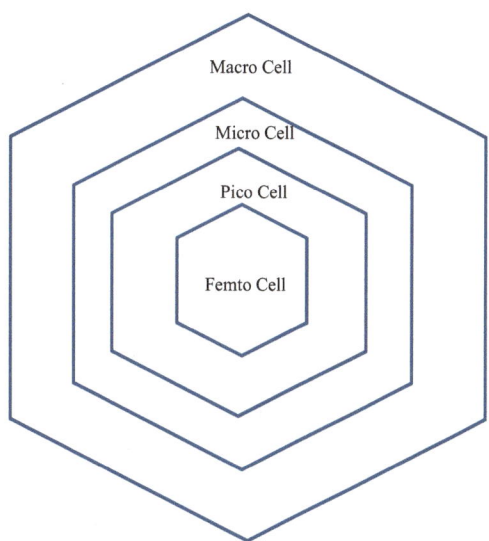

The break point depends on antenna location, antenna height, velocity of light and frequency. Therefore, the cell radius will vary accordingly.

7.4.2   *Microcell Deployment*

We envision that, Micro cells would be deployed in outdoor environment where the break point is given by the following equation: [5, 8]

$$R = d_o = \frac{4h_1 h_2 f}{c} \qquad (7.12)$$

Where,

R   Cell radius
d_o Fresnel zone break point ($R = d_o$)
h_1 Transmit antenna height
h_2 Receive antenna height
f   Frequency
c   velocity of light

Since R is proportional to the frequency and antenna height, and a Microcell needs to be smaller than a Macrocell, a lower operating frequency is required. For these reasons, use of frequencies lower than 1.9 GHz and appropriate antenna height, are recommended, yielding cell radii: $R \le 1$ km.

7.4.3   Picocell Deployment

In-building coverage is given by a three-ray model, where the antenna is located within the building having ground reflections as well as reflections from the ceiling. This is shown in Fig. 7.7. The break point is given by the following equation [5, 8]:

$$R = d_o = \frac{4(H - h_2)h_2 f}{c} \tag{7.13}$$

Where

$R = d_o$ Cell radius
H   Ceiling height
h_2 Receive antenna height
f   Frequency
c   Velocity of light

We also notice that the Picocell radius R is proportional to the frequency, and a Picocell needs to be smaller than a Microcell. Therefore a lower operating frequency could be used to design a Pico cell. For these reasons, use of frequencies lower than 1.9 GHz, such as 900 MHz, is recommended, yielding cell radii: $R \leq 0.5$ km. Note that using a higher frequency guarantees a free space propagation in an indoor environment.

7.4.4   Femtocell Deployment

In tunnels and subways, there exists a Fresnel Zone break point, where the direct path and the reflected path are exactly 180° out of phase. RF signals within this distance are in-phase and do not cancel each other out. On the other hand, RF signals beyond this point are out of phase and suffer from multipath cancellations. Therefore, by knowing the frequency, the transmit antenna can be properly positioned to create a free-space propagation in tunnels and subways as shown in Fig. 7.8.

The cell radius is given by the following equation:

$$R = d_o = \frac{4(D - h_1)(D - h_2)f}{c} \tag{7.14}$$

Where,

D   Diameter of the tunnel
h_1 Transmit antenna height
h_2 Receive antenna height
$R = d_o$ Cell radius
f   frequency
c   Velocity of light

Therefore, the cell radius in tunnels and subways is also proportional to the frequency. Since a Femto cell is smaller than Pico cell [9], a lower operating frequency could be used to design a Femto cell. For these reasons, use of frequencies lower than 900 MHz, such as 700 MHz, is recommended, yielding cell radii: R (micro) ≤ 0.2 km. Note that using a higher frequency guarantees a free space propagation in tunnels and subways.

7.5 Conclusions

- We have discussed Global RF & CO_2 Pollution connected to wireless communications.
- The classical Electron Spin Resonance (ESR) is presented to show that there is a possible
- public health issues due to RF absorption.
- It has also been argued that cell phone technology may contribute to global CO_2 pollution, expected to rise due to high speed data communication.
- With this in mind, we have presented a technique to design energy efficient green cellular technology, comprising Micro, Pico and Femto cells.

References

1. Saleh Faruque, "Global RF Pollution: A Distant Thunder":, Proceedings, IEEE-EiT 201,
2. Sumit Katiyar, Prof. R. K. Jain, Prof. N. K. Agrawal, "R.F. Pollution Reduction in Cellular Communication" in International Journal of Scientific & Engineering Research (IJSER), Vol. 3, Issue 3, March 2012.
3. Prof. R. K. Jain, Sumit Katiyar, Prof. N. K. Agrawal, "Hierarchical Cellular Structures in High-Capacity Cellular Communication Systems" in International Journal of Advanced Computer Science and Applications (IJACSA), Vol. 2, No 9, September 2011.
4. C. Kittel, Introduction to Solid State Physics", 3rd Edition, John Wiley & Sons, Inc., New York, 1966.
5. S. Faruque, "Cellular Mobile Systems Engineering", Artec House Inc., ISBN: 0-89006-518-7, 1996.
6. David R. Smith, "Digital Transmission Systems", Van Nostrand Reinhold Co. ISBN: 0442009178, 1985
7. Theodore S. Rappaport, "Wireless Communications", Pearson Education, ISBN: 81-7808-648-4, 2002.
8. Saleh Faruque,"A Three Ray Propagation Model for PCS and Micro cellular Services", IEEE MILCOM'95, Proceedings, pp. 1239–1243.
9. Analysys, "Picocells and Femtocells: Will indoor base-stations transform the telecoms industry?," [Online]. Available: http://research.analysys.com.